3D Printing in Mathematics

Proceedings of Symposia in
APPLIED MATHEMATICS

Volume 79

3D Printing in Mathematics

AMS Short Course
3D Printing: Challenges and Applications
January 3–4, 2022
Virtual

Maria Trnkova
Andrew Yarmola
Editors

AMERICAN MATHEMATICAL SOCIETY
Providence, Rhode Island

LECTURE NOTES PREPARED FOR THE AMERICAN MATHEMATICAL SOCIETY SHORT COURSE ON 3D PRINTING: CHALLENGES AND APPLICATIONS HELD VIRTUALLY JANUARY 3-4, 2022

The AMS Short Course Series is sponsored by the Society's Program Committee for National Meetings. The series is under the direction of the Short Course Subcommittee of the Program Committee for National Meetings.

2020 *Mathematics Subject Classification.* Primary 00A66, 97D40, 97N80, 97U60; Secondary 54C40, 14E20, 97I60, 14-04, 46E25, 20C20, 97D60.

For additional information and updates on this book, visit
www.ams.org/bookpages/psapm-79

Library of Congress Cataloging-in-Publication Data

Names: Trnkova, Maria, 1981- editor.
Title: 3D printing in mathematics / Maria Trnkova, Andrew Yarmola, editors.
Description: Providence, Rhode Island : American Mathematical Society, [2023] | Series: Proceedings of symposia in applied mathematics, 0160-7634 ; 79 | "2022 Virtual AMS Short Course on 3D Printing : Challenges and Applications, January 3-4, 2022"–Title page. | Includes bibliographical references and index.
Identifiers: LCCN 2023022844 | ISBN 9781470469160 (paperback) | ISBN 9781470474911 (ebook)
Subjects: LCSH: Three-dimensional imaging. | Three-dimensional imaging–Mathematics. | AMS: General – General and miscellaneous specific topics – Mathematics and visual arts, visualization. | Mathematics education – Education and instruction in mathematics – Teaching methods and classroom techniques. | Mathematics education – Numerical mathematics – Mathematical software, computer programs. | Mathematics education – Educational material and media, educational technology – Manipulative materials. | General topology – Maps and general types of spaces defined by maps – Algebraic properties of function spaces. | Algebraic geometry – Birational geometry – Coverings. | Mathematics education – Analysis – Functions of several variables. | Algebraic geometry – Explicit machine computation and programs. | Functional analysis – Linear function spaces and their duals – Rings and algebras of continuous, differentiable or analytic functions. | Group theory and generalizations – Representation theory of groups – Modular representations and characters. | Mathematics education – Education and instruction in mathematics – Student assessment, achievement control and rating.
Classification: LCC TA1560 .A18 2023 | DDC 621.9/880151–dc23/eng/20230601
LC record available at https://lccn.loc.gov/2023022844

10 9 8 7 6 5 4 3 2 1 28 27 26 25 24 23

Contents

Preface

Access to 3D printing facilities is quickly becoming ubiquitous across college campuses. However, while equipment training is readily available, the process of taking a mathematical idea and making it into a printable model presents a big hurdle for most mathematicians. Additionally, there are still many open questions around what objects are possible to print, how to design algorithms for doing so, and what kinds of geometries have desired kinematic properties. This volume is focused on the process and applications of 3D printing for mathematical education, research, and visualization, alongside a discussion of the challenges and open mathematical problems that arise in the design and algorithmic aspects of 3D printing.

The articles in this volume are focused on two main goals. The first one is to make a bridge between mathematical ideas and 3D visualization. The second is to describe methods and techniques for including 3D printing in mathematical education at different levels – from pedagogy to research and from demonstrations to individual projects. We hope to establish the groundwork for engaged academic discourse on the intersections between mathematics, 3D printing and education.

Traditional methods of manufacturing are centered around material removal, material defamation, casting, and joinery. This includes carving, milling, clay work, stamping metals or plastics and casting or molding. While advantageous in their choice of materials, many of these techniques suffer from issues of waste, precision, and the inability to make quick adjustments. Take, for example, Figure 1, which shows two wood carved polyhedra (a permutohedron and an associahedron). Production of these models requires a skilled carver who can get the correct angles and work carefully not to break any symmetry. Any mistake requires a fresh start. Getting a 3D printer to produce similar polyhedra is much faster and easier.

The advent of modern 3D printing technologies began in the early 1970s, while similar ideas and methods can be found in clay work dating back millennia. Dubbed additive manufacturing (AM), the process is very similar to integration: slice a 3D object by a foliation of $\mathbb{R}^3$ and use the thickened-up cross sections to approximate the object. The foliation in question is almost always by horizontal planes, but doesn't have to be, see [**2**]. Manufacturing then breaks down into "printing" each "layer" by *adding* material and fusing them together. The way these layers are created, fused, and supported define many aspects of the various technologies used in 3D printing. The most readily available version of this technology is a fused filament fabrication (FFF) printer, commonly also known as an FDM printer or extrusion printer. See Figure 2 for a diagram and description.

Current applications of AM include education, industry, dentistry and medicine (crowns, custom hearing aids, implants, prosthetics), footwear, architecture, aerospace

FIGURE 1. Wood carved associahedron and permutahedron

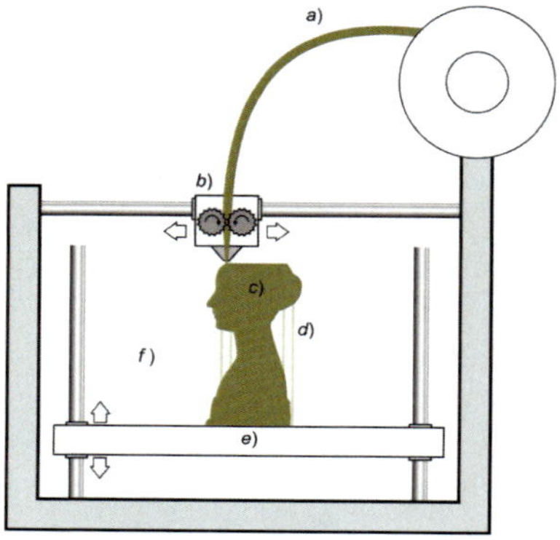

FIGURE 2. Diagram of an FFF printer: a filament (a) is fed through a moving nozzle (b) that melts, extrudes, and deposits material onto the print (c). A moving platform (e) lowers after each layer is deposited. Supports (d) are needed to temporarily support overhangs. Diagram reprinted with permission by Paolo Cignoni - Own work, CC BY-SA 4.0, `https://commons.wikimedia.org/w/index.php?curid=57953642` [**3**].

and home products. In addition, AM is considered to be more eco-friendly than traditional "subtractive" manufacturing because it uses material more efficiently. AM is a rapidly growing branch of engineering and it faces some challenges: mechanical and electrical engineering (time consuming, high cost, size limitations), material (rigidity, strength, magnetism, heat resistance, optical properties), software (geometric algorithms, complexity of interface and the design paradigms). For some of these technologies, for example, one must worry about the tool-path interfering with already printed aspects of the model (such as a printer nozzle colliding with the print during manufacture).

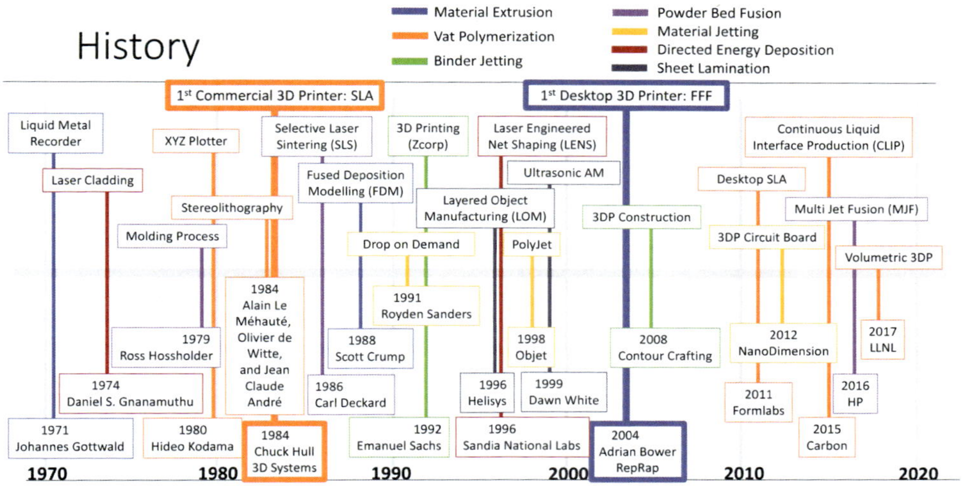

FIGURE 3. History of 3D printing technologies and inventions. Graphic by Bethany Weeks for the AMS Short Course 2022.

In Figure 3, we give an outline of the major events in the development of 3D printing technologies. Beginning with industrial application in the 70's, 80's and 90's, there is a renaissance in the mid-00's with the advent of desktop extrusion printers that now populate most university campuses. Industrial-scale processes are often more exact, produce finer detail, and a higher level of rigidity. Often such prints can be mail-ordered through an online service given a properly created 3D model for a reasonable price. In particular, access to a 3D printer is not necessary to participate in the field of mathematical visualization and design. Depending on ones needs, different printing material and methods should be considered. A table of the various printing technologies currently available along with their ISO/ASTM [**1**] definition is provided below.

Technology	**ISO/ASTM Process Description**
Material Extrusion	Material is selectively dispensed through a nozzle
Vat Polymerization	Liquid photopolymer in a vat is selectively cured by light-activated polymerization.
Binder Jetting	Liquid bonding agent is selectively deposited to join powder materials.
Powder Bed Fusion	Thermal energy selectively fuses regions of a powder bed.
Material Jetting	Droplets of feedstock material are selectively deposited.
Directed Energy Deposition	Focused thermal energy is used to fuse materials by melting as they are being deposited.
Sheet Lamination	Sheets of material are bonded to form a part.

Each technology has its strengths and weaknesses. These virtues and limits are too abundant to go into detail here, but we will give some examples. Material extrusion is generally a cheap, fast, and colorful method of printing, however it suffers from poor rigidity, low resolution, and the need for printed supports. It is one of the few technologies where intertwining various colors is relatively cheap and easy, which can be especially useful in education. Another popular technology is powder bed fusion, which is common for made-to-order services. It has the advantages of high detail, strength, uniformity, and does not require supports; yet it is rather slow and hazardous outside of industrial or lab environments due to the fine powders used. Lastly, we mention vat polymerization, which is available as a lab or desktop printer. This technology produces extremely high detail, smooth finish, and is affordable, but comes with the downsides of brittle prints, slow printing, and hazardous chemicals, requiring washing and curing cycles. Consideration should be taken when choosing printing technologies, but the complexities of running a printing lab can often be outsourced to your institution or a made-to-order service.

The works in this volume stem from the 2022 AMS Short Course on 3D printing. We would like to thank all of the speakers and participants that contributed to the short course and all of the authors that took their time to write contributing articles for this volume. In particular, we are grateful to our speakers silviana amethyst, David Bachman, Gabriel Dorfsman-Hopkins, Elisabetta Matsumoto, Henry Segerman, Timea Tihanyi, Bethany Weeks and volunteers Mark Ganter and Eric Moyer. A special thanks to Bethany Weeks for helping find 3D printing resources and volunteers locally in Seattle. We thank the AMS and its staff for helping run the course and in assembling this publication.

Maria Trnkova
Andrew Yarmola

Bibliography

[1] ISO/ASTM 52900:2021 *Additive manufacturing – General principles – Fundamentals and vocabulary*, International Organization for Standardization, Geneva, Switzerland, 2021 (en).

[2] Consuelo Rodriguez-Padilla, Enrique Cuan-Urquizo, Armando Roman-Flores, José L. Gordillo, and Carlos Vázquez-Hurtado, *Algorithm for the Conformal 3D Printing on Non-Planar Tessellated Surfaces: Applicability in Patterns and Lattices*, Applied Sciences **11** (2021), no. 16, 7509 (en).

[3] R. Scopigno, P. Cignoni, N. Pietroni, M. Callieri, and M. Dellepiane, *Digital Fabrication Techniques for Cultural Heritage: A Survey: Fabrication Techniques for Cultural Heritage*, Computer Graphics Forum **36** (2017), no. 1, 6–21 (en). DOI:10.1111/cgf.12781, via Wikimedia Commons, licensed under the Creative Commons Attribution-Share Alike 4.0 International license.

Proceedings of Symposia in Applied Mathematics
Volume **79**, 2023
https://doi.org/10.1090/psapm/079/00709

A mathematical overview and some applications of gear design

Elisabetta A. Matsumoto and Henry Segerman

Abstract. In this paper we give a brief overview of the geometry of involute gears, from a mathematical more than an engineering perspective. We also list some of the many variant geared mechanisms and discuss some of our 3D printed mechanisms.

1. Introduction

Gears are some of the most ubiquitous mechanical parts. They transfer torque between axles, often applying leverage as they do so, trading angular speed for torque and vice versa. Engineering texts on gears include many details that are used to characterize standardized commercial gears. In this paper, we aim to describe the construction of gears from a purely geometric perspective.

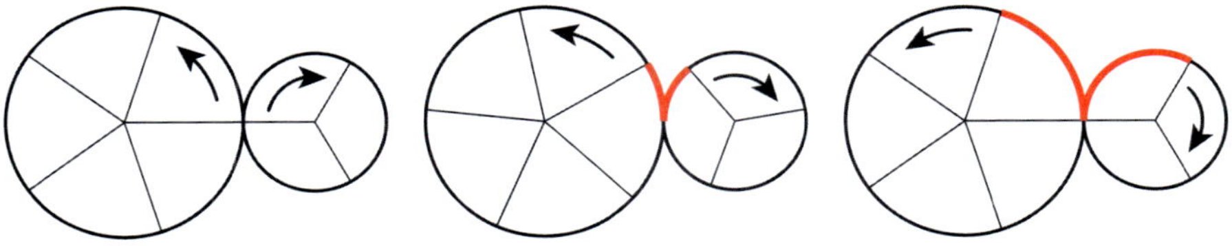

(A) (B) (C)

Figure 1.1. Rolling circles. In each diagram the red curves have equal length.

Consider two circles of radius r_1 and r_2 that rotate without slipping against each other, as shown in Figure 1.1. The fundamental principle of gears is that as the gears (or, for now, circles) rotate, the arc length swept out by the driving gear (with radius r_1, say) must be equal to the arc length swept out by the driven gear

2020 *Mathematics Subject Classification.* Primary 70B15, 00A66.

This material is based in part upon work supported by the National Science Foundation under Grant No. DMS-1439786 and the Alfred P. Sloan Foundation award G-2019-11406. The first author was supported in part by National Science Foundation grant DMR-1847172 and a Cottrell Scholars Award from the Research Corporation for Science Advancement. The second author was supported in part by National Science Foundation grant DMS-2203993.

(with radius r_2). For a circle of radius r that has rotated by angle θ radians, the arc length travelled is $s = r\theta$. Thus the angle that the driven gear rotates is

$$\theta_2 = (r_1/r_2)\theta_1$$

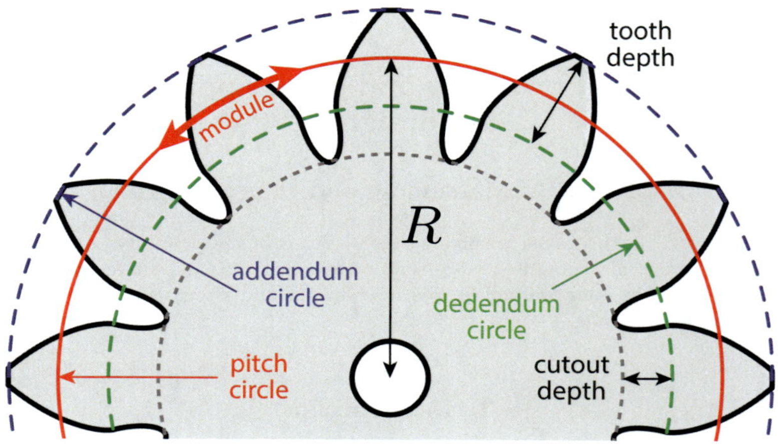

FIGURE 1.2. Anatomy of a gear.

We list some engineering terms that describe different parts of a gear, shown in Figure 1.2. These terms are not universally adopted, and other sources may vary. A pair of gears is based on two curves rotating around fixed axles that roll against each other without slipping. Such a curve, shown in red in the figure, is called the *pitch curve*. For now, we will consider circular gears, but in Sections 4.1 through 4.8 we will look at gears with acircular pitch curves. The rest of the terminology refers to the specifics of the gear teeth. The teeth are inscribed in a curve called the *addendum curve*, shown in blue. The bases of the teeth lie along the *dedendum curve* (sometimes called a *base circle*), shown in green. The distance between addendum and dedendum curves is the *tooth depth*. Some gears have an additional cutout between teeth for clearance, whose depth is called the *cutout depth*. The most important measurement is the *module* $m = R\frac{n}{2\pi}$, the fraction of the pitch curve of radius R from a gear with n teeth for each tooth.

2. Involute gears

2.1. Adding teeth to a gear. There are two main concerns when adding teeth to a gear. First we need to make sure that the teeth mes together, and second the gears should efficiently transfer torque from the driving gear to the driven gear. Teeth meshing makes the two gears rotate in registry – so that their respective pitch curves act as if they rotate without slipping. To ensure that the teeth in our gears mesh, we must have the same number of teeth per unit arc length on the two gears. So if our driving gear has n_1 teeth, our driven gear must have $n_2 = (r_1/r_2)n_1$ teeth. Of course both n_1 and n_2 must be integers. Frequently, gears that have not been precisely machined will use trapezoidal shaped teeth. This is practical for ensuring that teeth mesh, but the teeth won't contact each other smoothly, resulting in inefficient transmission of force.

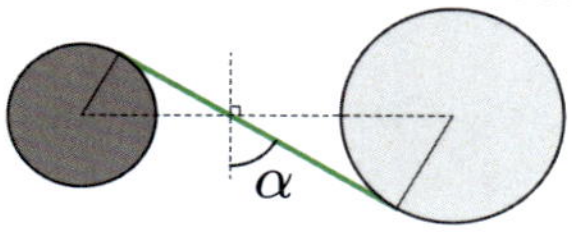

(A) Line of force (green) transmitted by gear teeth at pressure angle α.

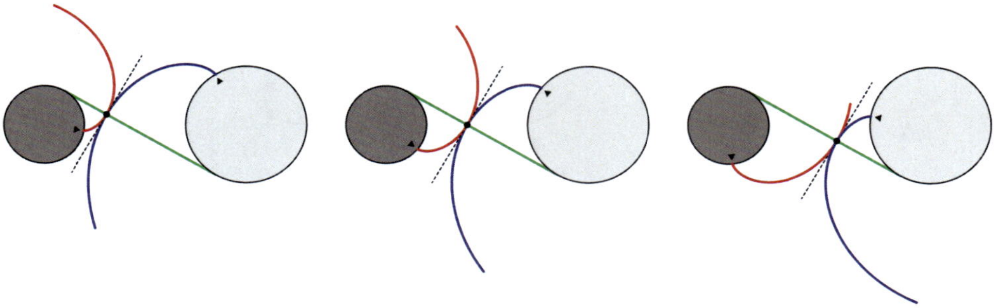

(B) Transmitting force along a line.

FIGURE 2.2. A pair of involutes based on circles of different radii rotating against each other. The dot shows the point of contact between the two curves.

2.3. Involutes. Almost all circular gears use *involutes* for the tooth flanks. (See Figure 2.4 for the construction. For an excellent animated introduction to involute gearing, see [**1**].) In physical gears this is critical for efficiency of torque transmission between components: throughout the rotation, we want equal torque to be transmitted from the driving gear to the driven gear.

For any pair of circles of radii r_1 and r_2 which form the base (or dedendum) circles for a gear pair, there exists a line tangent to both along which force is transmitted. In Figure 2.2b this line is drawn in green. As the red curve rotates clockwise, it applies force to the blue curve along the green line, causing it to rotate anticlockwise. As the gears rotate, the point of contact stays on the green line. Moreover, the shared line of tangency between the curves (dashed line) is always perpendicular to the line of contact. Together, these guarantee that the torque remains constant throughout the entire rotation of the gear.

The angle α that the green line makes with the line connecting the axles of the gears is called the *pressure angle*. See Figure 2.2a. (Most commercial gears use a pressure angle of $20°$, for its balance between force transmission and tooth strength.)

The red and blue curves of Figure 2.2b are involutes. Figure 2.4 shows a construction of an involute: we can think of it as the curve traced out by the end of a piece of string as it is wrapped around another curve $\gamma(t)$. The idea is to take a line of length ℓ that is tangent to γ at $t = t_0$ and draw a point. Then move (very slightly) down the curve γ by arc length a (at point $t = t_1$). Draw a point at the end of the line of length $\ell - a$ along the tangent at the point $\gamma(t_1)$. Repeat this process and connect the points. The result is an involute. If the curve γ is a circle parametrized by $\gamma(\theta) = \{r\cos\theta, r\sin\theta\}$ then the shape of its involute is given by

$$\boldsymbol{c}(\theta) = \{r(\cos\theta + \theta\sin\theta), r(\sin\theta - \theta\cos\theta)\}. \tag{2.5}$$

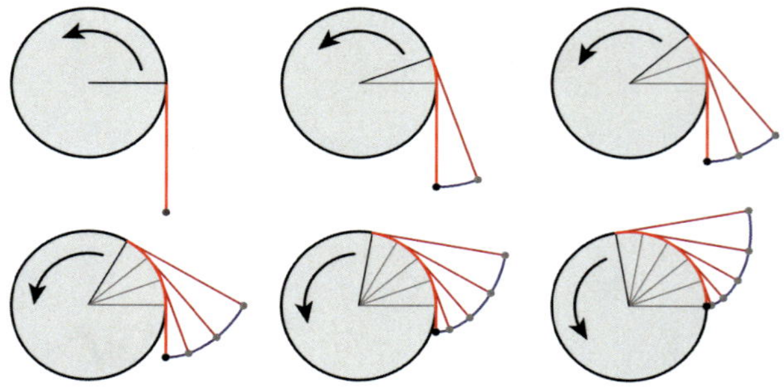

FIGURE 2.4. Construction of an involute curve.

More generally, the equation for an involute $c(t)$ to a curve $\gamma(t)$ at the parameter $t = t_0$ is:

$$\boldsymbol{c}_{t_0}(t) = \gamma(t) - \frac{\boldsymbol{c}(t)}{|\boldsymbol{c}(t)|}\int_{t_0}^{t} |\boldsymbol{c}(s)| ds. \tag{2.6}$$

2.7. Gear trains: module and pressure angle. The genius behind involute teeth lies not only in their ability to transmit force, but in the manufacturing process that machines them. In effect, gears are cut by other gears. The die used to cut the teeth on a gear is called a *rack*. This is a "straightened out gear". A rack can be seen as the limit of an involute gear as the radius of the pitch circle approaches infinity. The geometry of a rack is defined by the module m and the pressure angle α. See Figure 2.8.

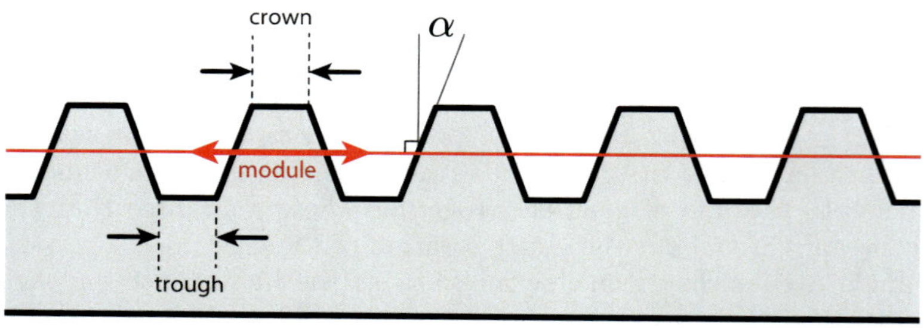

FIGURE 2.8. Schematic of a rack with pressure angle α.

In general, a pair of racks that are complementary (Figure 2.9a), can be used to cut a pair of gears (Figure 2.9b) that will mesh together (Figure 2.9c). In practice, gears are cut by racks that are self-complementary, where the tooth half-width is half the module, meaning that the crown and trough lengths of the rack are identical. Thus as long as two gears have pitch circles whose circumferences are an integer number of times the module of the generating die, then those two gears will mesh with each other. This technique can be used to make gear trains or other networks of multiple gears of different sizes.

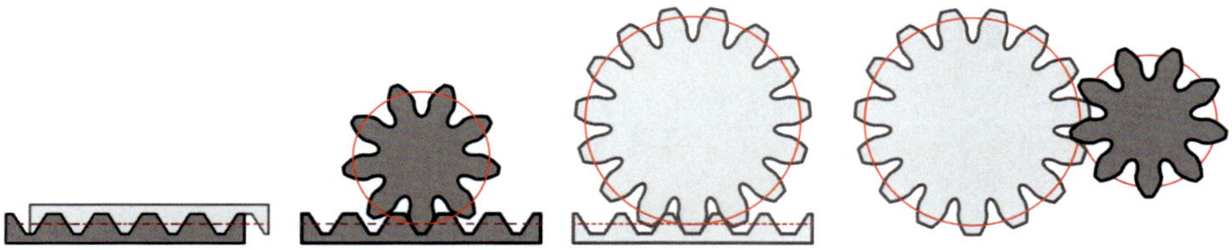

(A) Rack pair (B) Carving gears (C) Gear pair

FIGURE 2.9. Creating a pair of matching gears.

3. Traditional gear variants

Varying the construction given in Section 2 produces some closely related mechanisms. These variant contexts often require subtly different gear flank shapes to maintain optimal torque transfer. Giving parametric descriptions of these shapes can be difficult. In practice, engineers cut gears from dies to precisely match the complementary geometry, and therefore don't need parametric descriptions. We will not worry too much about these details and here just give a catalogue of different commonly used gear types.

3.1. Spur, helical, and herringbone gears. So far we have discussed essentially two-dimensional mechanisms, living in a plane. In real-world applications, the gears must be thickened up. The simplest way to thicken a planar gear is to extrude it perpendicular to the plane of the gear. (Mathematically, one would say that the three-dimensional shape is a product with an interval perpendicular to the plane.) This produces *spur gears*. If we twist the planar gear as we extrude then we get *helical gears*. If we glue two helical gears of opposite handedness together we get a *herringbone gear*. Herringbone gears have the advantage that a meshing pair tends to stay in position. See Figure 3.2.

FIGURE 3.2. From left to right: spur, helical, and herringbone gears.

3.3. Bevel gears. Spur, helical, and herringbone gears rotate around parallel axles that are perpendicular to the plane in which the gears are constructed. Moving the axes so that they meet at a point requires that the gears become *bevel gears*. Although not mathematically perfect, a good approximation to the correct shape for such a bevel gear can be formed by coning a planar involute gear to a point. (The correct shape seems to be a *spherical involute*, which is constructed analogously to a planar involute [**11**].) In our experience this approximation is good enough for 3D printed mechanisms. Figures 3.4 and 3.5 show two designs based on bevel gears (see [**9**] for a video). In Figure 3.4 the axles all meet at a single point at the center of a cube. In Figure 3.5, the axles meet at the vertices of an octahedron.

(A) (B)

FIGURE 3.4. Gear cube by the authors.

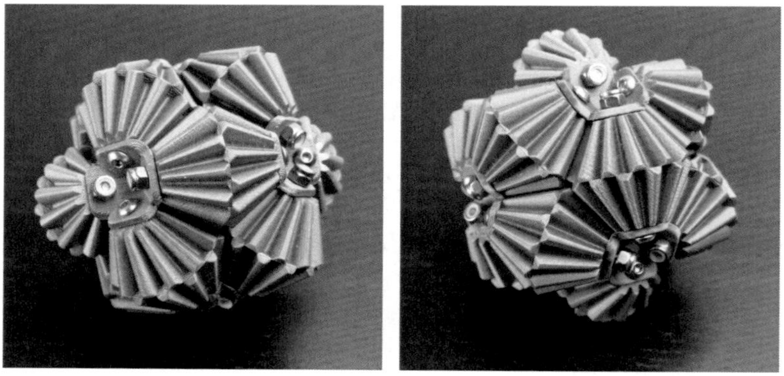

(A) (B)

FIGURE 3.5. Brain gear by the authors.

We have also used bevel gears to stabilize Buckminster Fuller's jitterbug mechanism, as shown in Figure 3.6. For more details, see [**7**].

3.7. Skew and worm gears. When the axles of a pair of meshing gears are neither parallel nor meet in a point, the gears become *skew gears*. A common mechanism of this kind is a *worm drive*, consisting of a worm (a skew gear similar to a screw) driving a *worm wheel*. This gives a large reduction ratio. See Figure 3.8.

4. Acircular gears

Acircular (or non-circular) gears are used in many engineering applications. Instead of having the pitch curve be a constant distance from the axle, acircular

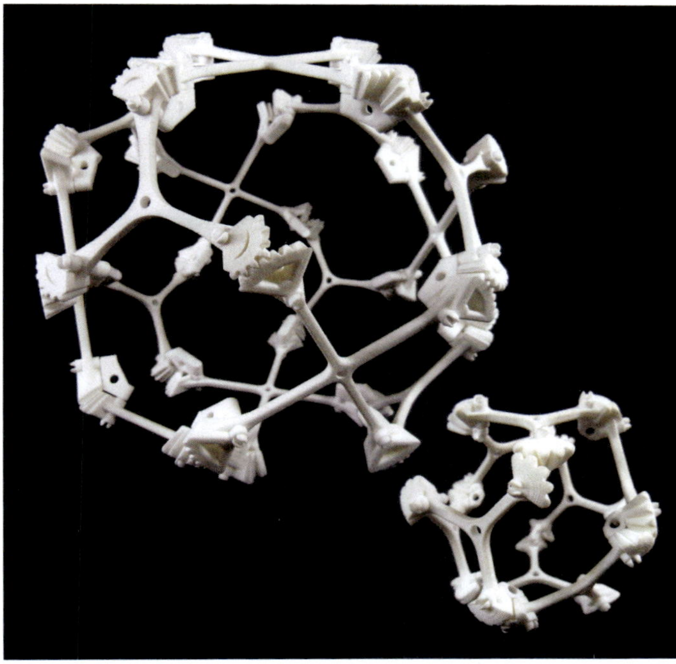

FIGURE 3.6. Geared jitterbugs by the authors. Videos on these designs: [**5**, **6**]

FIGURE 3.8. Skew gears (left) and a worm drive (right).

gears have a pitch curve $r(\theta)$ whose radius varies with the angle around the axle. The purpose of these gears is not to transmit torque with optimal efficiency but instead to control the gear ratio (as in a continuous transmission) or to control an additional motion (for example oscillations in a cam) [**2**].

4.1. Acircular pitch curves. We consider axles which are separated by a fixed distance, $a = r_1 + r_2$. If we know the pitch curve of the driving gear $r_1(\theta_1)$ then we can express r_2 in terms of θ_1. However, to make the driven gear, we need to express r_2 as a function of its own rotation coordinate, θ_2. Since the arclength swept out by each gear in time t must be the same, we end up with a relationship

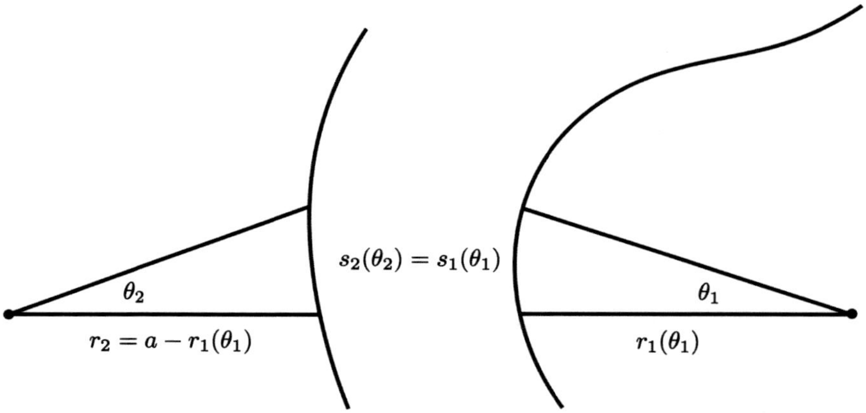

FIGURE 4.2. Geometry of an acircular gear pair.

between their radii and rotation rates, as illustrated in Figure 4.2:

$$r_1 d\theta_1 = r_2 d\theta_2 = \big(a - r_1(\theta_1)\big) d\theta_2 \tag{4.3}$$

If we assume that the driving gear with radius $r_1(\theta_1)$ rotates at a constant rate, then we can integrate Equation 4.3 to obtain the rotation rate of the driven gear:

$$\theta_2 + c = \int \frac{r_1(\theta_1)}{a - r_1(\theta_1)} d\theta_1.$$

This then enables us to calculate the pitch curve of the driven gear $r_2(\theta_2)$.

One classical example of an acircular gear pair is a nautilus gear pair, where the pitch curves are given by involutes of circles, as shown in Figure 2.2b.

4.4. Variable speed gears. For certain applications, we might not know the shape of the pitch curve of our acircular gears, but we do know the desired (variable) relative rotation rates between the two gears. That is, we require that the driven gear must rotate by some prescribed angle $\theta_2(\theta_1)$. An infinitesimal interpretation of this condition is given in Figure 4.5. Integrating this, we obtain

$$\int_0^{\theta_1} \sqrt{r_1(s)^2 + r_1'(s)^2} ds = \int_0^{\theta_2} \sqrt{r_2(t)^2 + r_2'(t)^2} dt.$$

This is unlikely to be analytically solvable. We give an iterative solution in **[7]**.

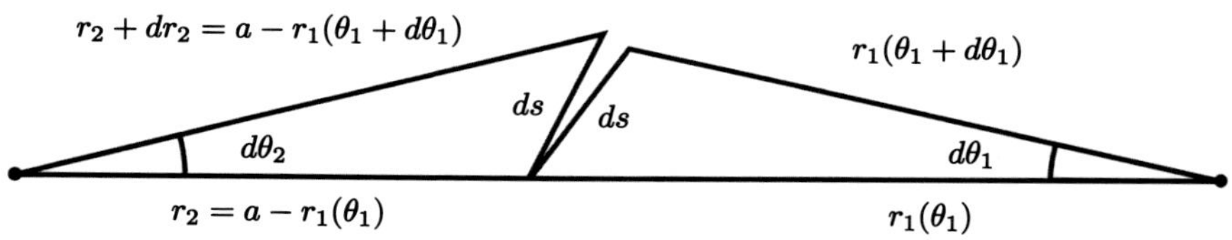

FIGURE 4.5. Differential elements of an acircular gear pair.

We needed to use variable speed gearing in the design of the larger jitterbug in Figure 3.6. This mechanism is based on a cuboctahedral shape. Each pair of bevel gears consists of a gear at a corner of a square face of the cuboctahedron meeting a gear at a corner of a triangular face. To get to the open state the square rotates by 90 degrees while the triangle rotates by 120 degrees. However, the rotation speeds of the two gears are not constant during this motion. For more details, again see **[7]**.

Note that both this construction and the construction given in Section 4.1 take advantage of the fact that rolling without slipping means that the arclength swept out by the two gears in a given time must be the same.

4.6. Cutting teeth on acircular gears. The process of adding teeth to acircular gears is quite similar to that discussed in Section 2.7. Again, the profile of the teeth is cut by a rack. The distance the rack travels as the gear rotates must be the same as the arclength swept out by the pitch curve during the motion. Additionally, the pitch curve in the rack must be tangent to the pitch curves of both gears. See Figure 4.7a. The envelope of the rack during this motion forms the shape of the teeth that need to be cut out. See Figure 4.7b. Since the pitch curve of the rack is not always perpendicular to the line connecting the two axles, the torque transmitted between the gears will not be constant during this motion. However this will produce teeth that have at least one point of contact at all times throughout the motion.

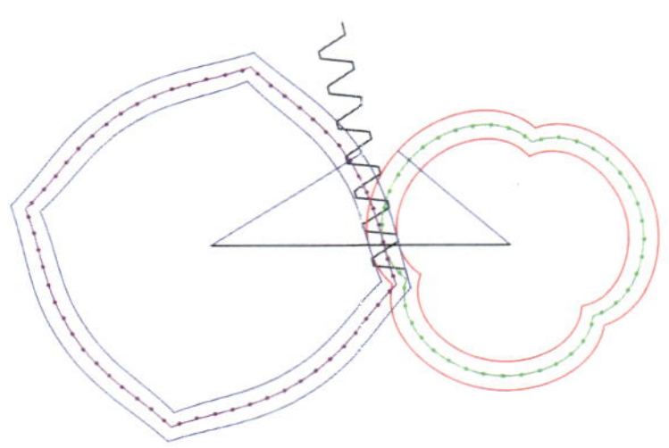

(A) The rack must be tangent to both pitch curves during cutting.

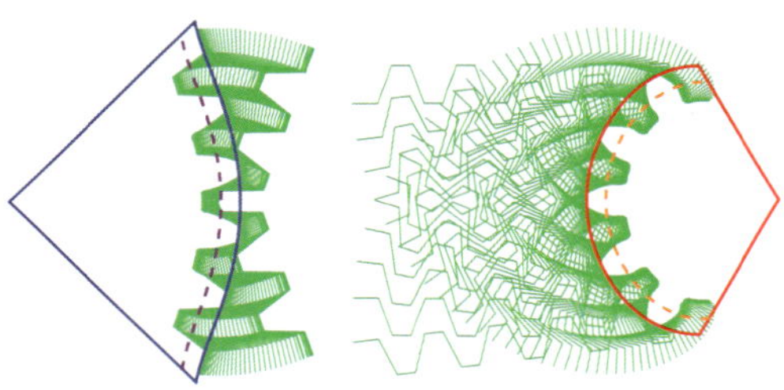

(B) The tooth shapes are given by the envelope of the rack as it sweeps along each gear.

FIGURE 4.7. Cutting teeth of acircular gears with a rack.

4.8. Alien gears. Alien gears are based on a pair of circular pitch curves with fixed radii r_1 and r_2. The (constant) rotation rates of the two gears are fixed by the ratio of their radii, but we take the shape of the "teeth" of the first gear to be some arbitrary shape, and must determine a corresponding shape for the second gear. Conceptually, this can be achieved by carving the shape of the second gear from a large block of material by rotating the gears against each other at their constant rotation rates. This is similar to the situation in Section 4.6, but with the rack replaced by the first gear. In more detail, we begin with an arbitrary shape, S_1 and two gear centers, shown in Figure 4.9a. We put ourselves in the frame of reference of the generated gear. The combined motion of the two rotations has S_1 roll around the second center, c_2, as shown in Figure 4.9b. Figure 4.9c shows the full collection of positions for S_1 as it rolls around c_2. Assuming that the union of all of these positions has a hole in the middle containing c_2, we take S_2 to be the shape of the hole.

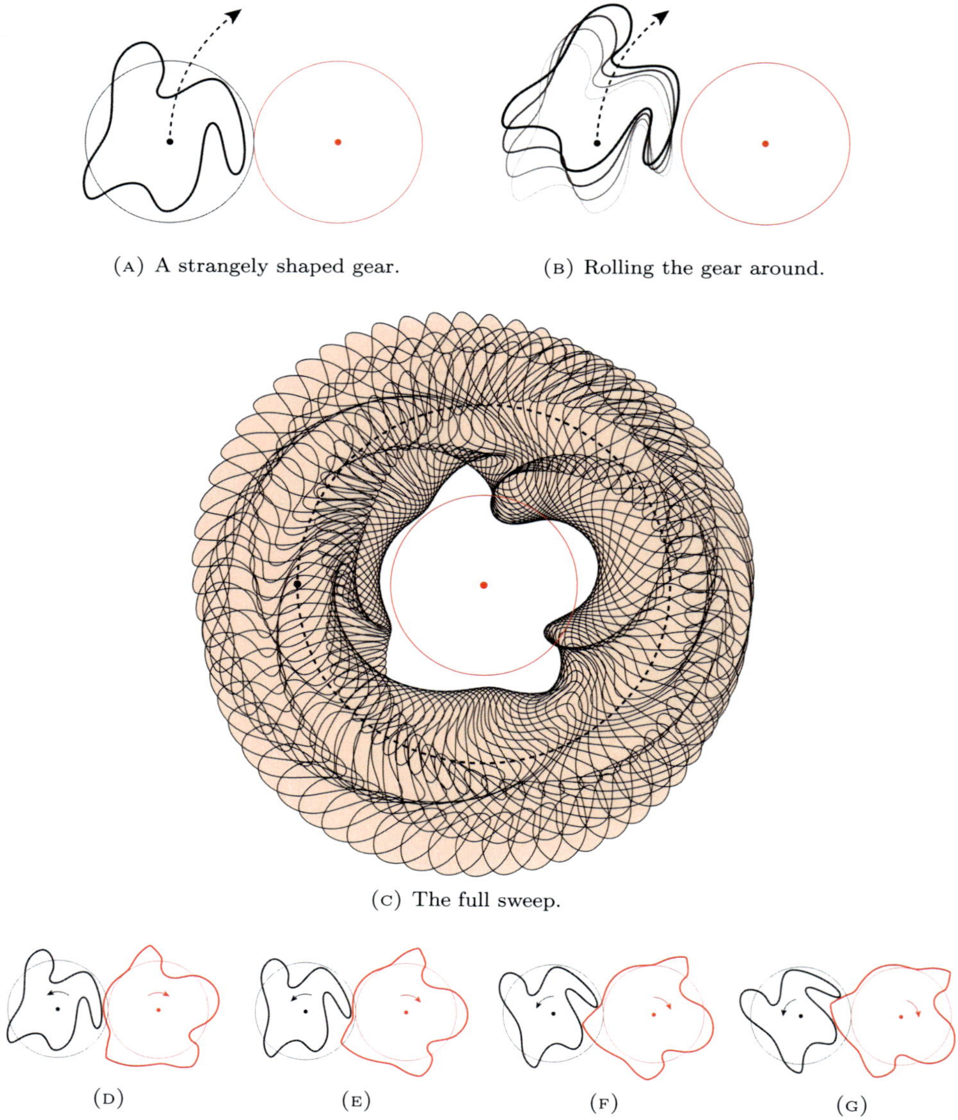

(A) A strangely shaped gear.

(B) Rolling the gear around.

(C) The full sweep.

(D)

(E)

(F)

(G)

FIGURE 4.9. Carving alien gears.

In Figures 4.9d to 4.9g we show the resulting gear shapes rotating against each other. Depending on the choice of shape S_1, torque transfer will likely not be optimal – indeed, the driving gear may not even force the driven gear to turn, and so the gears may slip against each other.

REMARK 4.1. We made Figure 4.9 using a Rhino-Grasshopper script by Arkuo Zheng. Rather than continuously sweeping S_1 around c_2, this uses a finite collection of positions for S_1. It also smooths the shape of the hole in the center of Figure 4.9c to produce the gear shown in Figures 4.9d to 4.9g. For simple shapes and if we don't need much precision, this is good enough. Various approaches to computing continuous sweeps have been proposed. See [**3**, Section 2] for an overview of work in this area.

5. Applications

In this section, we describe some of our projects in 3D printed geared mechanisms.

5.1. Odd numbers of gears. Images of gears are often used in graphic design to give a sense of parts working together for a shared goal. In graphic design, the gears do not need to actually work – in particular it is surprisingly common to see a design with an odd number of spur gears meshing in a loop. Adjacent spur gears rotate in opposite directions, so an odd number of these arranged in a loop is frozen in place.

An interesting challenge then is to construct a loop with an odd number of (necessarily) non-spur gears. Our first solution was *Triple gear*, see Figure 5.2a. Here the three rings rotate around skew axes, so are skew gears. Unusually, some of the gearing surfaces face in towards the axle of the gear. See [**15**] for more details. A significantly simpler solution uses three helical gears, as shown in Figure 5.2b.

(A) Triple gear by Saul Schleimer and the second author. Videos on this design: [**12, 14**]

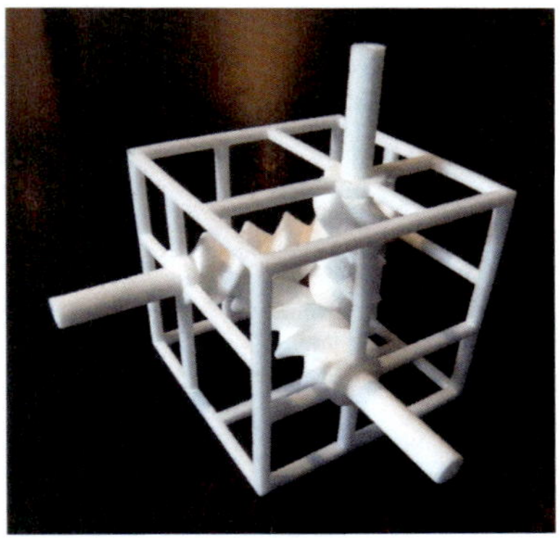

(B) *Triple helix* by Saul Schleimer and the second author. Video on this design: [**16**]

FIGURE 5.2.

5.3. Sliding racks. Racks allow a mechanism to convert the traditional rotational movement of a gear into translation. Mechanisms with gears and racks are usually designed to avoid parts sliding against each other. Sometimes however, sliding is the only way to achieve a desired outcome. Figure 5.4 shows some of our mechanisms that use racks sliding against each other. The first of these, *Borromean racks*, was also inspired by the odd-gear challenge.

5.5. Gripping gears. Another challenge we set ourselves was to make a pair of gears that mesh and rotate together, be unable to separate, but have no axles attached to an outer frame. Our *gripping gears* (joint work with Will Segerman) give a solution to this. See Figure 5.6.

(A) *Borromean racks* (B) *Tetrahedral racks* (C) *Five axis racks*

FIGURE 5.4. Rack designs. The first two designs here are by Saul Schleimer and the second author, the third is by the authors. Videos on these designs: [**4**, **13**, **17**]

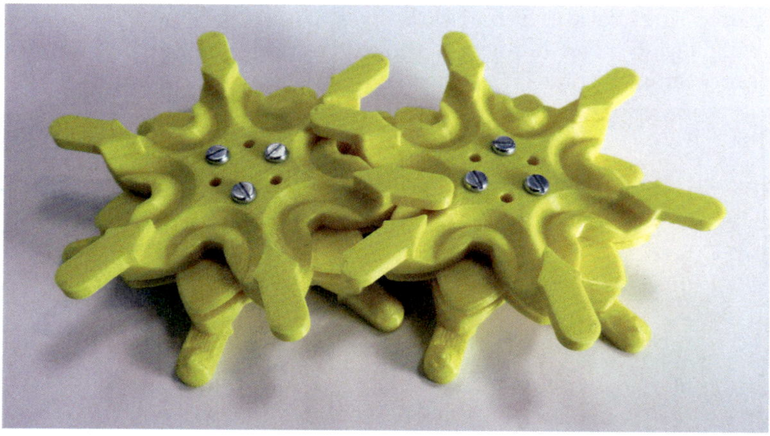

FIGURE 5.6. *Gripping gears*. Video on this design: [**10**]

Arms extending from each gear terminate in pegs which fit into grooves on the other gear. While a peg is in a groove, it restricts the movement of the two gears relative to each other. With enough such pegs and grooves, the two gears should be restricted to only be able to rotate around each other in the desired way.

Figure 5.7 shows a schematic diagram of two pegs in their corresponding grooves. The shape of the groove is an *epitrochoid*: the curve traced by a point attached to a circle as it rolls around a second, fixed circle. See Figure 5.8.

If trapped between two parallel planes, the two gears shown in Figure 5.7, remain connected together for most of the desired motion. However, there is a configuration in which the two parts can be slid apart. In Figure 5.7, if we slide the gear on the left further to the left and slightly up, then it can be disentangled from the gear on the right. Even worse, if the two gears are allowed to twist out of the plane then the pegs disengage and they can be untangled easily. To keep the parts together even when twisting is allowed, we make each part from two offset copies of the gear, sandwiched around two interleaving extra layers. The two copies

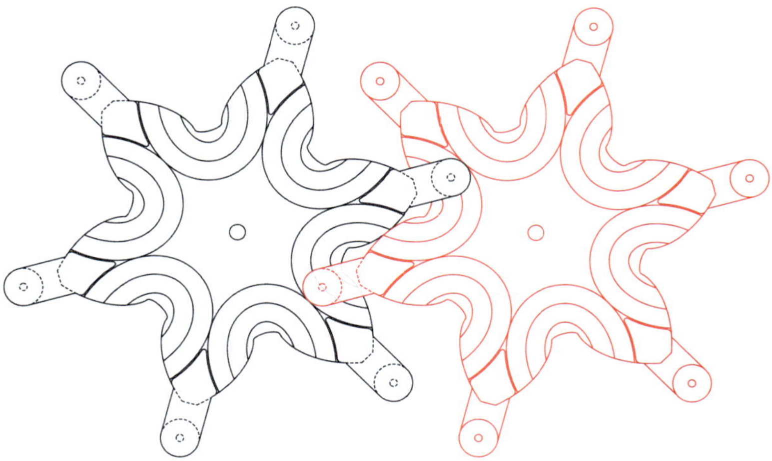

FIGURE 5.7. Gripping gears schematic diagram.

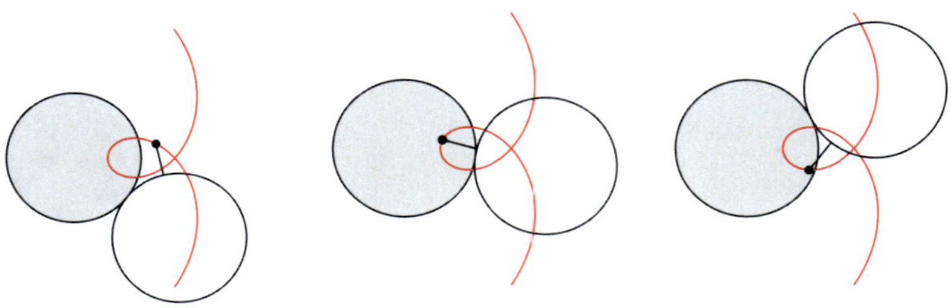

(A) (B) (C)

FIGURE 5.8. Three frames of an animation: the white circle rolls around the outside of the grey circle. The end of an arm attached to the white circle traces out an epitrochoid in the frame of reference of the grey circle.

protect each other's "blind spots" for planar disengagement, while the interleaving extra layers prevent twisting out of plane. This then adds enough constraints to allow only the desired motion.

In fact, there are enough constraints that we can remove some material and still have the mechanism work. In Figure 5.9 we have tunneled holes through some of the involute interfaces between the two gears. This makes the gear movement slightly less smooth but otherwise has little effect, and allows a solid rod (in Figure 5.9b a screwdriver) to pass between the two gears.

5.10. Braiding gears. One of our motivations in designing gripping gears was as a precursor to gears that would mesh and hold together for some part of their motion, but then disengage at some point. Our *braiding gears* achieve this.

In Figure 5.11a, three identical gears are arranged in a line, with the outer two gears meshing only with the center gear. In Figure 5.11b and Figure 5.11c the outer two gears rotate around below the center gear, until in Figure 5.11d they meet, and the configuration has three-fold rotational symmetry. (In fact, it has

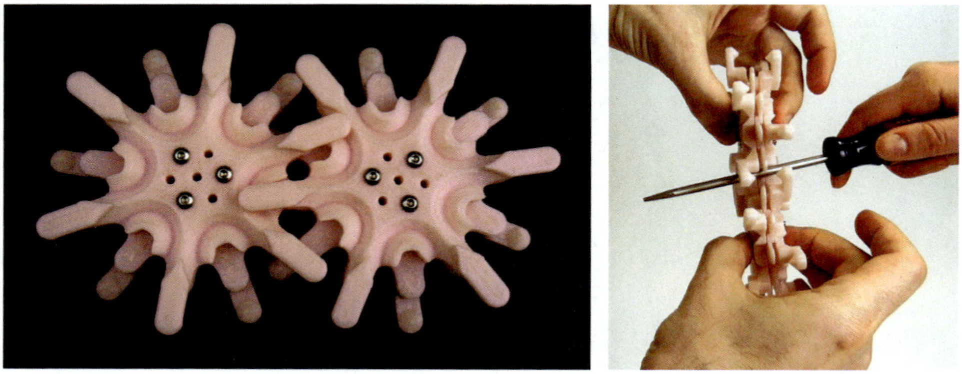

(A) (B)

FIGURE 5.9. Gripping gears with holes. Video on this design: [**8**]

(A)

(B) (C) (D)

FIGURE 5.11. *Braiding gears*. Video on this design: [**18**]

dihedral symmetry.) Thus, we may move away from this symmetric state in three different ways, each resulting in a different gear becoming the center gear in the line configuration. The video **[18]** shows this motion.

(A) (B)

FIGURE 5.12. Braiding gears schematic diagram.

Figure 5.12a shows the upper side of two gears in the line configuration. Note that here they are translates of each other, and that their engaged pegs are in epicycloid grooves, just as for gripping gears. Rotating the left gear by $\pi/3$ anti-clockwise and rotating the right gear by $\pi/3$ clockwise results in Figure 5.12b. The grooves in use here are made from two parts, meeting at a corner. The first part is a subarc of the standard epicycloid from the gripping gear design: a peg moves along this part as we rotate the gears from their positions in Figure 5.12a to those in Figure 5.12b. The second part is determined by the motion of the two gears that become outermost as we move from the three-fold symmetric configuration in Figure 5.11d back to that in Figure 5.11a: it is the path followed by a peg on one of those two gears as it moves relative to the other gear.

FIGURE 5.13. Four levels.

These grooves and the corresponding pegs then allow the motion shown in Figure 5.11. However, as shown in Figure 5.13, the design has parts extending beyond and above the grooves and pegs. Moving from bottom to top, there are a total of four "levels" to the design: level zero has the grooves in it, level one are the arms below which the pegs hang. Levels two and three are there to solve the following problem.

Since the configuration in Figure 5.11d has three-fold symmetry, we can break the symmetry in three different ways. In particular, since the lower right gear could become the central gear, it is possible for the lower left gear to detach from the upper gear. If the lower right gear were not there to "catch" the lower left gear, then it would detach from the other two gears, and the mechanism would fall apart.

Thus, it is necessary that when in the line configuration, if one of the two outer gears rotates down then the other must also rotate down. In other words, if one outer gear rotates counterclockwise, then the other must rotate clockwise. This is the purpose of levels two and three – on the two outer gears these levels interfere with each other, forcing the outer gears to stay in phase.

We would have preferred if the two outer gears smoothly geared into each other. Unfortunately there is a lot of traffic moving around at the different levels as the mechanism moves, which restricts the size and shape of the upper level arms, and seems to preclude the possibility of perfectly smooth motion.

As in the gripping gears, the gears shown in Figure 5.12 can be slid apart at particular points of their motion. Again, we solve this problem by bolting together two copies of the design, back to back. Here, the extra arms introduced on levels two and three have the side benefit of preventing twisting motions, and so the interleaving layers are not needed.

Acknowledgments

Part of this material was completed while the authors were in residence at the Institute for Computational and Experimental Research in Mathematics in Providence, RI, during the Illustrating Mathematics program.

References

[1] Gergely Bencsik, *Involute gears explained*, July 2022, https://www.youtube.com/watch?v=nrsCoQN6V4M.

[2] Bálint Laczik, *Involute profile of non-circular gears*, Available from http://manuals.chudov.com/Non-Circular-Gears.pdf.

[3] Jinesh Machchhar, Henry Segerman, and Gershon Elber, *Conjugate shape simplification via precise algebraic planar sweeps toward gear design*, Computers & Graphics **90** (2020), 1–10.

[4] Elisabetta A. Matsumoto and Henry Segerman, *Five axis racks*, September 2018, https://www.youtube.com/watch?v=dkguSyeQXjc.

[5] Elisabetta A. Matsumoto and Henry Segerman, *Geared jitterbug*, October 2018, https://www.youtube.com/watch?v=kwERR5flAOU.

[6] Elisabetta A. Matsumoto and Henry Segerman, *Geared cuboctahedral jitterbug*, March 2019, https://www.youtube.com/watch?v=eZFgHnvtPUo.

[7] Elisabetta A. Matsumoto and Henry Segerman, *Geared jitterbugs*, Proceedings of Bridges 2019: Mathematics, Art, Music, Architecture, Education, Culture (Phoenix, Arizona) (Susan Goldstine, Douglas McKenna, and Kristóf Fenyvesi, eds.), Tessellations Publishing, 2019, pp. 399–402.

[8] Elisabetta A. Matsumoto and Henry Segerman, *Gripping gears with pass-through holes*, October 2019, https://www.youtube.com/watch?v=RBZG8M8_a8Y.

[9] Elisabetta A. Matsumoto and Henry Segerman, *Gear cube and Brain gear*, December 2022, https://www.youtube.com/watch?v=l4eL1MvEbN0.

[10] Elisabetta A. Matsumoto, Henry Segerman, and Will Segerman, *Gripping gears*, September 2019, https://www.youtube.com/watch?v=ENFXnNtd3xU.

[11] Noh Gill Park and Hyoung Woo Lee, *The spherical involute bevel gear: its geometry, kinematic behavior and standardization*, Journal of Mechanical Science and Technology **25** (2011), no. 4, 1023–1034.

[12] Saul Schleimer and Henry Segerman, *Triple gear*, December 2012, https://www.youtube.com/watch?v=I9IBQVHFeQs.
[13] Saul Schleimer and Henry Segerman, *Borromean hairpins*, September 2013, https://www.youtube.com/watch?v=WQ9ptuUxfk4.
[14] Saul Schleimer and Henry Segerman, *Powered triple gear*, April 2013, https://www.youtube.com/watch?v=QhXjevOY_uk.
[15] Saul Schleimer and Henry Segerman, *Triple gear*, Proceedings of Bridges 2013: Mathematics, Music, Art, Architecture, Culture (Phoenix, Arizona) (George W. Hart and Reza Sarhangi, eds.), Tessellations Publishing, 2013, pp. 353–360.
[16] Saul Schleimer and Henry Segerman, *Triple helix*, March 2013, https://www.youtube.com/watch?v=74ygvXzmrgk.
[17] Saul Schleimer and Henry Segerman, *Tetrahedral racks*, August 2016, https://www.youtube.com/watch?v=sDJk-9QxFro.
[18] Henry Segerman, *Braiding gears*, November 2019, https://www.youtube.com/watch?v=Lh7yAbw0H24.

School of Physics, Georgia Institute of Technology, Atlanta, Georgia 30332; and International Institute for Sustainability with Knotted Chiral Meta Matter, Hiroshima University, Higashihiroshima, Japan
Email address: sabetta@gatech.edu

Department of Mathematics, Oklahoma State University, Stillwater, Oklahoma 74078
Email address: henry.segerman@okstate.edu

Proceedings of Symposia in Applied Mathematics
Volume **79**, 2023
https://doi.org/10.1090/psapm/079/00710

Mathematics of floating 3D printed objects

Daniel M. Anderson, Brandon G. Barreto-Rosa, Joshua D. Calvano, Lujain Nsair, and Evelyn Sander

ABSTRACT. We explore the stability of floating objects through mathematical modeling and experimentation. Our models are based on standard ideas of center of gravity, center of buoyancy, and Archimedes' Principle. We investigate a variety of floating shapes with two-dimensional cross sections and identify analytically and/or computationally a potential energy landscape that helps identify stable and unstable floating orientations. We compare our analyses and computations to experiments on floating objects designed and created through 3D printing. In addition to our results, we provide code for testing the floating configurations for new shapes, as well as giving details of the methods for 3D printing the objects. The paper includes conjectures and open problems for further study.

CONTENTS

1. Introduction

Interest in the dynamics of icebergs has been driven by the desire to describe various natural phenomena (including their rolling) as well as practical considerations associated with shipping and protection of offshore structures in arctic environments. Allaire's [**Al**] study of iceberg stability was motivated by the need to assess mitigation strategies, such as towing of icebergs, to reduce threats posed by icebergs to offshore structures. Allaire identified readily-identifiable above-water characteristics, such as the ratio of waterline width to above-water height, to estimate

2020 *Mathematics Subject Classification.* Primary 54C40, 14E20; Secondary 46E25, 20C20.
Key words and phrases. Center of gravity, center of buoyancy, Archimedes' Principle.

stability for a menagerie of iceberg shapes – Blocky, Drydock, Dome, Pinnacled, Tabular, Growler. Bailey [**B**], motivated by similar concerns, examined stability of icebergs in terms of rolling frequency. The potential for iceberg stability considerations to be dynamic even in calm water due to underwater melting/dissolution was considered by Deriabyn & Hjorth [**DH**]. They were also interested to identify what practical, above-water, observations could be made to predict stability changes driven by underwater changes in iceberg morphology.

Ship design and the design of other man-made floating objects has no doubt driven much scientific and technical work in this field. We make no attempt to review this literature but interested readers may find resources in the work of Mégel & Kliava [**MK**] and Wilczynski & Diehl [**WD**]. Historically speaking, scientific thought on this dates back to Archimedes (c. 287–212/211 B.C.). See for example Rorres [**Ro**], a fascinating article on the original work of Archimedes and extensions thereof.

Whereas icebergs and ships are complex three dimensional floating objects, the floating objects that are the focus of the present work are those whose configurations can effectively be characterized in two-dimensions. Specifically, we shall consider 'long' objects whose cross sections are constant for the full length of the object. Such shapes have been the focus of popular online Apps, such as Iceberger [**Ice1**] and the remixed version [**Ice2**]. These effective two-dimensional floating objects have proven to be mathematically tractable yet rich in observable phenomena. Prediction of the stable orientations for a long beam with square-cross section and uniform density, for example, was considered by Reid [**Re**]. Depending on the ratio of the density of the object to the density of the fluid any rotation of the square can be a stable floating orientation (i.e. any orientation from flat side up to corner up). This configuration has been recently revisited both experimentally and analytically by Feigel and Fuzailov [**FF**]. These authors validated experimentally that for a small range of density ratios near 0.25 (and also similarly near 0.75) the full range of stable orientations can be realized. Another view of this which we discuss in more detail in our work is that owing to the four-fold symmetry of the square, there can be either four or eight stable orientations of the floating square depending on the density ratio. We explore how these results change for what we denote as 'off-center' squares.

The case of a long 'floating plank' of rectangular cross section was the focus of work by Delbourgo [**D**]. Here, in addition to the density ratio, another parameter – the aspect ratio of the rectangle – appears. Delbourgo identified in this density ratio vs. aspect ratio space the existence of six characteristic floating configurations in terms of (1) long side up or short side up, (2) top side parallel or not parallel to the waterline, and (2) the number of submerged vertices. Further studies have explored other cross sectional shapes and investigated details of the breaking of left-right symmetries of the floating shapes as the density ratio is varied (e.g. see Erdös, Schibler, & Herndon [**E1**] for square and equilateral triangle cross sections and Erdös, Schibler, & Herndon [**E2**] for the three-dimensional shapes of the cube, octahedron, and decahedron). Work in this area has focused on homogeneous objects with uniform density. We shall relax this assumption in our present work to include some special classes of non-uniform density in which the center of gravity of the square no longer resides at its centroid (cf. Definition 2.1).

FIGURE 1. The floating Mason M.

An excellent review of many important mathematical ideas – Archimedes' Principle, Center of Gravity, Center of Buoyancy, and the notion of Metacenter – related to floating objects is the work of Gilbert [**G**]. One particularly useful concept is that of the potential energy of floating objects. Gilbert shows that if one can compute the potential energy landscape as a function of all possible orientations of the object one can identify stable floating configurations by the locations of local minima of the potential energy function. This is one of the main objectives of our computations as we then use this to identify stable orientations.

The present work shares some of the spirit of the paper by Feigel and Fuzailov [**FF**] to revisit these questions both theoretically and experimentally. Our experiments, however, are conducted with 3D printed shapes. While some of our theoretical effort has been on objects with square cross section, our methodology is motivated by the recognition that 3D printing offers the opportunity to float objects whose cross sections are effectively limited only by ones own creativity in defining new shapes. An example shape we analyze is the 'Mason M' shown in Figure 1.

The new contributions we make here are (1) predicting floating configurations of an object with square cross section when the center of gravity is not at the object's centroid, (2) predicting floating configurations for objects of general two-dimensional polygonal cross section, and (3) verifying the predictions of (1) and (2) as well as classical ones for objects of square cross section with experiments conducted with 3D printed objects.

This paper is organized as follows. In Section 2, we introduce definitions and terminology, outline prior results for long floating objects of uniform density with square cross section, and state our analytical results for the case of long floating objects with square cross section with non-uniform density. In Section 3 we give a detailed description of how to 3D print floating objects. This section contains sufficient details and accompanying code so that interested readers would be able to print their own objects and perform their own experiments. Note that we have additionally provided codes and files in a GitHub repository [**GIT**]. In Section 4 we describe how we obtain experimental results on our 3D printed floating objects. In Section 5 we describe our code for computing stable floating configurations for objects with general polygonal cross sections. In Section 6 we describe the results of our floating experiments and relate them to our theoretical results for several cases involving square cross sections and a selected case of a nontrivial cross section. In Section 7 we give our conclusions and discuss some open problems.

2. Mathematical models for floating objects

2.1. Definitions. In this section, we give some basic definitions and concepts that will be needed for the remainder of our analysis.

DEFINITION 2.1 (Center of gravity). If the mass distribution is given by a continuous density function of $\rho(x,y,z)$ within a domain Ω, then the center of gravity can be obtained by

$$(G_x, G_y, G_z) = \frac{1}{M_{\text{obj}}} \iiint_\Omega (x,y,z)\, \rho(x,y,z)\, dV \ ,$$

where M_{obj} is the object's mass. In the case of uniform density, i.e. ρ is a constant independent of x, y and z, the center of gravity is called the centroid.

For a long object of length L with uniform cross section, with uniform density in the long direction, i.e. $\rho(x,y,z) = \rho(x,y)$ is independent of z, the center of gravity is given by

$$\vec{G} = (G_x, G_y) = \frac{L}{M_{\text{obj}}} \iint_\Omega (x,y)\rho(x,y)\, dA$$

with $G_z = L/2$ (relative to one end of the object).

LEMMA 2.2. *For an object of length L with a polygonal cross-sectionwith uniform constant density ρ, we can compute the area and center of gravity as sums involving only the vertices of the polygon. In particular, let*

$$\{(x_1, y_1), \ldots, (x_N, y_N), (x_1, y_1)\}$$

be the vertices of the cross section polygon, oriented counterclockwise. Then the mass of the object is

$$M_{\text{obj}} = \rho L A \ ,$$

where the area A of the polygon of cross section is given by

$$A = \frac{1}{2} \sum_{k=1}^{N} (x_k + x_{k+1})(y_{k+1} - y_k) \ , \tag{2.1}$$

a result also known as the shoelace formula, and the center of gravity $\vec{G} = (G_x, G_y)$ is given by

$$\begin{aligned} G_x &= \frac{1}{6A} \sum_{k=1}^{N} (x_k^2 + x_k x_{k+1} + x_{k+1}^2)(y_{k+1} - y_k) \\ G_y &= \frac{1}{6A} \sum_{k=1}^{N} -(y_k^2 + y_k y_{k+1} + y_{k+1}^2)(x_{k+1} - x_k) \, . \end{aligned} \tag{2.2}$$

PROOF. According to Green's theorem

$$\iint_\Omega \left(\frac{dg}{dx} - \frac{df}{dy}\right) dA = \oint_{d\Omega} f\,dx + g\,dy,$$

Choose $(f,g) = (0,x)$ so that $\frac{dg}{dx} - \frac{df}{dy} = 1$. We can parametrize the line segment between (x_k, y_k) and (x_{k+1}, y_{k+1}) by $(x,y) = (1-t)(x_k,y_k) + t(x_{k+1},y_{k+1})$ where $t \in (0,1)$. For this line segment, we get $dx = (x_{k+1} - x_k)dt$ and $dy = (y_{k+1} - y_k)dt$. Now we evaluate the integral from Green's theorem:

$$\begin{aligned}\int_0^1 x\,dy &= \int_0^1 (x_k + t(x_{k+1} - x_k))(y_{k+1} - y_k)\,dt\\ &= x_k(y_{k+1} - y_k) + \frac{1}{2}(x_{k+1} - x_k)(y_{k+1} - y_k)\\ &= \frac{1}{2}(x_{k+1} + x_k)(y_{k+1} - y_k)\,.\end{aligned}$$

Now we use this to calculate the full area:

$$A = \iint_\Omega dA = \oint_{d\Omega} x\,dy = \frac{1}{2}\sum_{k=1}^{N}(x_{k+1} + x_k)(y_{k+1} - y_k)\,.$$

Now we turn our attention to the calculation of G_x. We use Green's theorem but this time with $(f,g) = (0, x^2/2)$ in order to satisfy $\frac{dg}{dx} - \frac{df}{dy} = x$. Again for the line segment from (x_k, y_k) to (x_{k+1}, y_{k+1}) we get

$$\begin{aligned}\iint_\Omega x dA &= \oint_{d\Omega} \frac{x^2}{2}\,dy\\ &= \frac{1}{2}\int_0^1 (x_k + t(x_{k+1} - x_k))^2(y_{k+1} - y_k)\,dt\\ &= \frac{1}{6}(x_k^2 + x_k x_{k+1} + x_{k+1}^2)(y_{k+1} - y_k)\,.\end{aligned}$$

Since ρ is constant, we get that G_x is equal to $\rho L/M_{\text{obj}}$ times this integral, but $M_{\text{obj}} = \rho L A$, and this gives the factor of $1/A$ written in the formula above. In a similar way, using Green's theorem with $(f,g) = (-y^2/2, 0)$, we get

$$\iint_\Omega y dA = -\frac{1}{6}(y_k^2 + y_k y_{k+1} + y_{k+1}^2)(x_{k+1} - x_k)\,.$$

This gives the equation for G_y stated above. □

The proof above relies on Green's theorem, but there are other ways to derive this formula, such as dividing the object into triangles or trapezoids, and combining the corresponding triangle or trapezoid area and center of gravity formulas.

DEFINITION 2.3 (Buoyancy). Buoyancy is a force exerted on an object that is wholly or partially submerged in a fluid. The magnitude of this force is equal to the weight of the displaced fluid. Buoyancy relates to the density of the fluid, the volume of the displaced fluid, and the gravitational field; it is independent of the mass and density of the immersed object. The buoyancy force acts vertically upward at the centroid of the displaced volume. The *center of buoyancy* is given by

$$(B_x, B_y, B_z) = \frac{1}{V_{\text{sub}}}\iiint_{\Omega_{\text{sub}}} (x,y,z)\,dV, \tag{2.3}$$

where V_{sub} is the submerged volume of the object, and Ω_{sub} is the submerged domain. Like in the case of center of gravity, in the uniform cross section case $B_z = L/2$. We use the notation $\vec{B} = (B_x, B_y)$. Note that $\vec{B}$ is the centroid of the submerged domain.

We now state one of the major results necessary for understanding floating objects.

THEOREM 2.4 (Archimedes' principle). *The upward buoyant force exerted on an object wholly or partially submerged is equal to the weight of the displaced fluid. In the absence of other forces, such as surface tension, this can be expressed in the force balance as*

$$M_{\text{obj}} g = \rho_{\text{f}} V_{\text{sub}} g, \tag{2.4}$$

where g is acceleration due to gravity, ρ_{f} is the density of the fluid, and V_{sub} is the submerged volume of the object.

Observe that (2.4) represents a balanced (net zero) equation of competing forces with the left terms representing gravitational force and right term representing the opposing force of buoyancy. Note that if the object has uniform density, then the mass of the object can be written as $M_{\text{obj}} = \rho_{\text{obj}} V_{\text{obj}}$. In this case, it follows that

$$\frac{V_{\text{sub}}}{V_{\text{obj}}} = \frac{\rho_{\text{obj}}}{\rho_{\text{f}}}. \tag{2.5}$$

For our purposes, Archimedes' Principle determines the appropriate waterline intersections defining a submerged volume whose value relative to the total volume matches the appropriate density ratio. However, it is important to note that satisfying Archimedes' Principle is not a sufficient condition for determining a stable equilibrium. An equilibrium orientation of a floating body occurs when the center of gravity and the center of buoyancy are vertically aligned. If $\vec{G}$ lies directly below $\vec{B}$ the equilibrium is stable, whereas if $\vec{G}$ lies above $\vec{B}$ the equilibrium may or may not be stable. We present an alternative approach using energy principles similar to that of Erdös [**E1**] and Gilbert [**G**]. After identifying a waterline that is consistent with Archimedes' Principle, we define a unit vector normal to the waterline, by keeping the object fixed and rotating the frame of reference (waterline) by angle θ to generate all orientations satisfying Archimedes' Principle. The stable positions of a floating body occur at the minima of the potential energy. The potential energy function for a floating body is given by

$$U(\theta) = \hat{n}(\theta) \cdot (\vec{G} - \vec{B}(\theta)), \tag{2.6}$$

where $\hat{n}$ is the unit normal vector to the waterline pointing out of the water, and θ is the rotation angle of the waterline. Note that $\vec{B}(\theta)$ and $\hat{n}(\theta)$ depend on θ, but $\vec{G}$ is independent of θ. Below, we derive formulas for the stable floating configurations by finding minima for $U(\theta)$.

2.2. Square Cross Section Revisited. The stability of a long floating object with square cross section, and the corresponding nontrivial floating configurations, have been investigated theoretically in a number of studies. Reid [**Re**] provided the first theoretical identification of stable floating equilibrium configurations based on arguments using forces and moments. Feigel & Fuzailov [**FF**] provided a recent alternative derivation of these equilibrium conditions, a brief review of related studies, and also detailed experiments validating the theory. Their experiments had a particular focus on floating configurations in the transition from 'flat side up' orientations to 'corner up' orientations. We revisit the square cross section configuration here with the goal of writing down the entire potential energy landscape, whose minima reveal the stable equilibrium configurations. The square has four-fold symmetry, and we exploit this in the identification of a center of buoyancy formula. A new contribution we make in the present work corresponds to situations in which the center of gravity is not at the center of the square. We specifically explore the breaking of this four-fold symmetry for floating objects with square cross sections and use the corresponding potential energy landscapes to understand the observations. In sections that follow, we demonstrate that the identification of potential energy landscapes can be obtained for shapes of more general cross sections in order to understand their stable floating configurations.

Rather than fix a waterline and consider different orientations of the square we fix a reference frame on the square with corners at $(1,-1)$, $(1,1)$, $(-1,1)$, $(-1,-1)$ and consider different orientations of the waterline. Three configurations are relevant as shown in Figure 2 – the first has the waterline intersecting opposite sides of the square and the second and third have the waterline intersecting adjacent sides of the square. We work out these three cases below and then give the generalization for all orientations.

For our $[-1,1]^2$ square, the cross sectional area is $A_{\text{obj}} = 4$. If we denote by A_{sub} the submerged area, Archimedes' principle requires that

$$\frac{A_{\text{sub}}}{A_{\text{obj}}} = R, \tag{2.7}$$

where $R \in (0,1)$ is the density ratio $\rho_{\text{obj}}/\rho_{\text{f}}$ of the floating object to the fluid. We shall assume that the object's density is uniform throughout but if it were not the appropriate interpretation of ρ_{obj} for the application of Archimedes' principle would be the effective density – i.e. the object's mass divided by the volume of the object. Note that in the present context we work in terms of cross sectional area; corresponding volumes would be obtained by multiplying the cross sectional area by the length of the object in the third dimension.

Below we outline the computation of the center of buoyancy, $\vec{B}(\theta)$, as a function of orientation θ for the two cases in which (1) the waterline intersects opposite sides of the square and (2) the waterline intersects adjacent sides of the square.

2.2.1. *Waterline intersects opposite sides of square.* Here we define the waterline by the equation

$$y \quad = \quad x\tan\theta + H, \tag{2.8}$$

where θ is the slope of the waterline and H is the y-intercept (see Figure 2). With the water assumed to occupy the region below the waterline, the submerged area can be written in terms of H as $A_{\text{sub}} = 2(1+H)$. Therefore, Archimedes' Principle requires $R = (1+H)/2$, or equivalently $H = 2R-1$. Note that for $R \in (0,1)$ it follows that $H \in (-1,1)$.

We define waterline intersection points $(-1, y_L)$ and $(1, y_R)$ and note that

$$y_L = -\tan\theta + H, \quad y_R = \tan\theta + H. \tag{2.9}$$

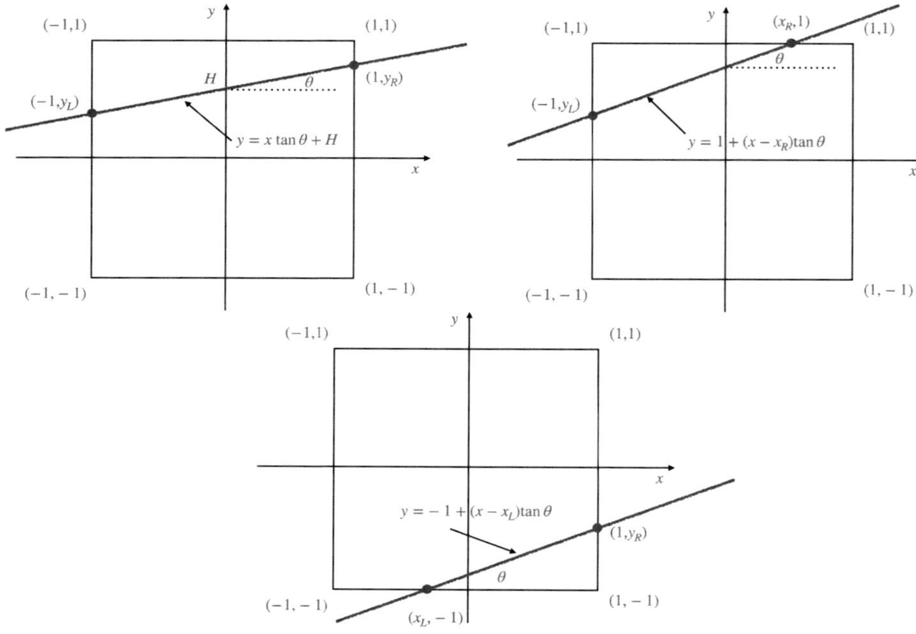

FIGURE 2. The sketch on the upper left shows the configuration in which the waterline (blue) intersects opposite sides of the square. The sketch on the upper right shows the configuration in which the waterline intersects adjacent sides of the square for $R > 1/2$. The sketch on the bottom shows the configuration in which the waterline intersects adjacent sides of the square for $R < 1/2$.

By definition, the configuration under consideration requires that $y_L \in [-1, 1]$ and $y_R \in [-1, 1]$. Furthermore, the largest and smallest θ occur for $y_R = \pm 1, y_L = \mp 1$, meaning that $-\pi/4 \leq \theta \leq \pi/4$ and $-1 \leq \tan\theta \leq 1$. Combining these facts with the definitions of y_L and y_R, we get

$$\tan\theta + H \leq 1 \quad \text{and} \quad -\tan\theta + H \leq 1 \quad \text{if } H \geq 0, \tag{2.10}$$

$$-\tan\theta + H \geq -1 \quad \text{and} \quad \tan\theta + H \geq -1 \quad \text{if } H \leq 0, \tag{2.11}$$

These can be rewritten as

$$-1 + |H| \leq \tan\theta \leq 1 - |H|. \tag{2.12}$$

This range of $\tan\theta$ corresponds to a range of θ values $[\theta_1^{\min}, \theta_1^{\max}]$ contained in $[-\pi/4, \pi/4]$. By symmetry, there is a corresponding configuration when rotated by $\pm\pi/2$ and $\pm\pi$.

The submerged area in this configuration is defined by the four points $(1, -1)$, $(1, y_R)$, $(-1, y_L)$, and $(-1, -1)$. We use (2.2) to find the center of buoyancy as the centroid of the submerged boundary region. In particular, let (x_k, y_k) be given by

$\{(1,-1),(1,y_R),(-1,y_L),(-1,-1),(1,-1)\}$. Then $\vec{B}(\theta)=(B_x(\theta),B_y(\theta))$ where

$$B_x(\theta)=\frac{1}{6A_{\text{sub}}}\sum_{k=1}^{4}(x_k^2+x_kx_{k+1}+x_{k+1}^2)(y_{k+1}-y_k)$$

$$B_y(\theta)=\frac{1}{6A_{\text{sub}}}\sum_{k=1}^{4}-(y_k^2+y_ky_{k+1}+y_{k+1}^2)(x_{k+1}-x_k)\,.$$

Computing these sums, combined with the values of y_L and y_R and the fact that $A_{\text{sub}}=2(1+H)$, we find that the center of buoyancy takes the form

$$\vec{B}(\theta)=\vec{B}_1(\theta)\equiv\frac{1}{2(1+H)}\Big(\frac{2}{3}\tan\theta,-1+H^2+\frac{1}{3}\tan^2\theta\Big), \tag{2.13}$$

where θ can take on any value defined by the inequalities (2.12). For use below we define this specific form for the center of buoyancy as $\vec{B}_1(\theta)$.

2.2.2. *Waterline intersects adjacent sides of square.* A similar approach can be applied to the second configuration shown in the upper right sketch of Figure 2. Here we give expressions for the results when $R\geq 1/2$ and when $R<1/2$.

Case 1: $R\geq 1/2$. Here we assume that the waterline intersects the left and top sides of the square at points $(-1,y_L)$ and $(x_R,1)$ so that three corners of the square are submerged. We consider the case $R<1/2$ in the next section although note that this case can be carefully extracted from the present case (according to Gilbert [**G**], Feigel & Fuzailov [**FF**], among others).

Here we identify the waterline by

$$y \;=\; (x-x_R)\tan\theta+1, \tag{2.14}$$

where $y_L=-(1+x_R)\tan\theta+1$.

This waterline cuts a triangular region of area $\frac{1}{2}(1+x_R)(1-y_L)$ from the original square. This means that $A_{\text{sub}}=4-\frac{1}{2}(1+x_R)(1-y_L)$ and

$$R=\frac{A_{\text{sub}}}{A_{\text{obj}}} \;=\; \frac{4-\frac{1}{2}(1+x_R)(1-y_L)}{4}=\frac{4-\frac{1}{2}(1+x_R)^2\tan\theta}{4}. \tag{2.15}$$

Rearranging this gives x_R in terms of R and the waterline slope $\tan\theta$

$$(1+x_R)^2 \;=\; \frac{8-8R}{\tan\theta}. \tag{2.16}$$

Conditions on θ come from the requirement that $0\leq(1+x_R)\leq 2$ and $-1\leq y_L\leq 1$. The first of these reveals that

$$0\leq\frac{2-2R}{\tan\theta}\leq 1. \tag{2.17}$$

For the case under consideration $0<\theta<\frac{\pi}{2}$. It follows that $\tan\theta\geq 2-2R$. Equality corresponds to the waterline passing through the point $(1,1)$ and $(-1,y_L)$. The other extreme corresponds to the waterline passing through the point $(-1,-1)$ and $(x_R,1)$. This has $\tan\theta=2/(1+x_R)$. Here the triangular area is $\frac{1}{2}2(1+x_R)$ which means $R=(4-(1+x_R))/4$ or $(1+x_R)=4-4R$. Since θ cannot exceed this angle we have $\tan\theta<2/(4-4R)$. Thus, for this configuration the value of $\tan\theta$ is constrained by

$$2-2R\leq\tan\theta\leq\frac{1}{2-2R}. \tag{2.18}$$

As in the previous case, the center of buoyancy is obtained by calculating the centroid of the submerged area using (2.2). In particular, using the counterclockwise oriented vertices of the submerged polygon:

$$\{(x_R, 1), (-1, y_L), (-1, -1), (1, -1), (1, 1)\},$$

we can calculate the following integral

$$\vec{B}(\theta) = \frac{1}{A_{\text{sub}}} \iint_{\Omega_{\text{sub}}} (x, y)\, dA \tag{2.19}$$

as a sum. Evaluating this integral gives

$$\begin{aligned} \vec{B}(\theta) = \vec{B}_2^+(\theta) \equiv \frac{1}{A_{\text{sub}}}\Big(1 - \frac{1}{2}(y_L + 1) - \frac{1}{6}(x_R^3 + 1)\tan\theta, \\ -1 + \frac{1}{2}(1 - x_R) + \frac{(1 - y_L^3)}{6\tan\theta}\Big). \end{aligned} \tag{2.20}$$

For use below, we define this specific form for the center of buoyancy as $\vec{B}_2^+(\theta)$. Recall that $A_{\text{sub}} = 4R$ and

$$y_L = -\tan\theta(1 + x_R) + 1, \quad (1 + x_R)^2 = \frac{8 - 8R}{\tan\theta}, \tag{2.21}$$

where $\tan\theta$ satisfies (2.18). This condition gives a range of θ values given by $[\theta_2^{+\,\min}, \theta_2^{+\,\max}]$ contained in $[0, \pi/2]$. Again by symmetry, we get a corresponding set of angles by adding $\pm\pi/2, \pm\pi$.

Case 2: $R < 1/2$. Here assume that the waterline intersects the square at points $(x_L, -1)$ and $(1, y_R)$ so that only the lower right corner of the square is submerged.

Here we identify the waterline by

$$y = \tan\theta(x - x_L) - 1, \tag{2.22}$$

where $y_R = \tan\theta(1 - x_L) - 1$.

This waterline cuts a triangular region of area $A_{sub} = \frac{1}{2}(1 + y_R)(1 - x_L)$ from the original square. This means that

$$R = \frac{A_{\text{sub}}}{A_{\text{obj}}} = \frac{1}{8}(1 + y_R)(1 - x_L) = \frac{(1 - x_L)^2 \tan\theta}{8}. \tag{2.23}$$

Rearranging this gives x_L in terms of R and the waterline slope $\tan\theta$

$$(1 - x_L)^2 = \frac{8R}{\tan\theta}. \tag{2.24}$$

Conditions on θ come from the requirement that $0 \le (1 - x_L) \le 2$. This translates to

$$\tan\theta \ge 2R. \tag{2.25}$$

Also the condition $-1 \le y_R \le 1$ leads to

$$0 \le \tan\theta \le \frac{2}{1 - x_L}. \tag{2.26}$$

Using $8R = (1 - x_L)(1 + y_R)$ leads to $0 < \tan\theta < 1/(2R)$.

So, together these require

$$2R \le \tan\theta \le \frac{1}{2R}. \tag{2.27}$$

As before, the center of buoyancy satisfies

$$\vec{B}(\theta) \quad = \quad \frac{1}{A_{\text{sub}}}\int_{\Omega_{\text{sub}}} (x,y)\, dA, \tag{2.28}$$

which can be calculated as a sum involving the vertices of the submerged polygonal cross section via (2.2). It follows that

$$\begin{aligned} \vec{B}(\theta) &= \vec{B}_2^-(\theta) \\ &\equiv \frac{1}{A_{\text{sub}}}\left(\frac{y_R+1}{2} - \frac{(1-x_L^3)}{6}\tan\theta, -\frac{(1-x_L)}{2} + \frac{1+y_R^3}{6\tan\theta}\right). \end{aligned} \tag{2.29}$$

For use below, we define this specific form for the center of buoyancy as $\vec{B}_2^-(\theta)$. Recall that $A_{\text{sub}} = 4R$ and

$$y_R = \tan\theta(1-x_L) - 1, \quad (1-x_L)^2 \quad = \quad \frac{8R}{\tan\theta}, \tag{2.30}$$

where $\tan\theta$ satisfies (2.27).

2.2.3. *Potential energy expressions: square cross section.* As defined in (2.6) the potential energy function is given by

$$U(\theta) = \hat{n}(\theta)\cdot(\vec{G} - \vec{B}(\theta)),$$

where the unit normal to the waterline can be expressed as a function of θ as $\hat{n}(\theta) = (-\sin\theta, \cos\theta)$. For the square defined above with uniform density the center of gravity $\vec{G} = (0,0)$. However, we are interested in a generalization of the square where the center of gravity, by some means, is not necessarily located at the center but rather has coordinates $\vec{G} = (G_x, G_y)$. Note that as long as Archimedes' principle is applied with the appropriate mass of the object, the calculations presented above for the center of buoyancy are independent of the location of the center of gravity. So, in what follows we treat $\vec{G}$ as nonzero in general.

For $R \geq 1/2$ define

$$U_{B_1}(\theta) \quad = \quad \hat{n}(\theta)\cdot\vec{B}_1(\theta)\,, \text{ for } -1+|H| \leq \tan\theta \leq 1-|H|\,, \tag{2.31}$$

which corresponds to a range of $\theta \in [\theta_1^{\min}, \theta_1^{\max}]$ defined by (2.12). Also define

$$U_{B_2^+}(\theta) \quad = \quad \hat{n}(\theta)\cdot\vec{B}_2^+(\theta)\,, \text{ for } 2-2R \leq \tan\theta \leq \frac{1}{2-2R}\,, \tag{2.32}$$

which corresponds to a range of $\theta \in [\theta_2^{+\min}, \theta_2^{+\max}]$ defined by (2.18).

We can write the potential energy function $U(\theta)$ as follows

$$U(\theta) \quad = \quad \begin{cases} \hat{n}(\theta)\cdot\vec{G} - U_{B_1}(\theta) & \theta \in [\theta_1^{\min}, \theta_1^{\max}] \\ \hat{n}(\theta)\cdot\vec{G} - U_{B_1}(\theta \pm \frac{\pi}{2}) & \theta \pm \frac{\pi}{2} \in [\theta_1^{\min}, \theta_1^{\max}] \\ \hat{n}(\theta)\cdot\vec{G} - U_{B_1}(\theta \pm \pi) & \theta \pm \pi \in [\theta_1^{\min}, \theta_1^{\max}] \\ & \\ \hat{n}(\theta)\cdot\vec{G} - U_{B_2^+}(\theta) & \theta \in [\theta_2^{+\min}, \theta_2^{+\max}] \\ \hat{n}(\theta)\cdot\vec{G} - U_{B_2^+}(\theta \pm \frac{\pi}{2}) & \theta \pm \frac{\pi}{2} \in [\theta_2^{+\min}, \theta_2^{+\max}] \\ \hat{n}(\theta)\cdot\vec{G} - U_{B_2^+}(\theta \pm \pi) & \theta \pm \pi \in [\theta_2^{+\min}, \theta_2^{+\max}] \end{cases} \tag{2.33}$$

A similar formula applies when $R < 1/2$ (replace B_2^+ with B_2^- and the corresponding range for $\tan\theta$ given in (2.27)).

2.3. Squares with off-center weights. We also explore the case of a square cross section with an off-center weight parallel to the long axis of the object. Specifically we consider 3D printed objects with a hole in the square that can be filled with a material of different density. In our experiments we had the option to leave the hole as void space or to insert a nail cut to fit the object. In either case, before floating the object tape was placed over the holes to prevent water from filling the space.

For such a configuration we can predict the modified center of gravity $\vec{G} \neq 0$. In particular, consider the same square with corners at $(1,-1)$, $(1,1)$, $(-1,1)$, $(-1,-1)$ with a hole with circular cross section at point (a,b) with $a \in (0,1)$, $b \in (0,1)$, and radius r_H. When such an object is printed there is a border around the hole whose thickness we denote by t and whose density is ρ_{PLA} (i.e. the density of the solid print material). With the hole filled with a nail whose density is ρ_{nail} we can compute the center of gravity of the object as a whole (printed object plus nail) as

$$M_{\text{obj}}\vec{G} = L\left\{\int_{\Omega_0} \rho(\vec{x})\vec{x}\,dA + \int_{\Omega^{\text{hole+border}}} \rho(\vec{x})\vec{x}\,dA\right\}, \tag{2.34}$$

where M_{obj} is the mass of the whole object (including the nail if one is inserted), L is the length of the object, Ω_0 denotes the cross-sectional domain of the square excluding the hole and border and $\Omega^{\text{hole+border}}$ denotes the circular cross section that includes the (printed) border of the hole and the hole, and $\rho(\vec{x})$ denotes the material density at position $\vec{x}$ in the plane. The square printed without a hole will have some void space in its interior and this can be controlled by changing the infill of the print. In our squares printed with a hole this infill region gets replaced by the hole plus the border material of the hole. Therefore, it is convenient to rewrite the formula (2.34) for center of gravity $\vec{G}$ as

$$\begin{aligned}
\frac{M_{\text{obj}}}{L}\vec{G} &= \int_{\Omega_0} \rho(\vec{x})\,\vec{x}\,dA + \int_{\Omega^{\text{hole+border}}} \rho(\vec{x})\,\vec{x}\,dA \\
&\quad + \int_{\Omega^{\text{hole+border}}} \rho_{\text{infill}}\,\vec{x}\,dA - \int_{\Omega^{\text{hole+border}}} \rho_{\text{infill}}\,\vec{x}\,dA, \\
&= \int_{\Omega} \rho(\vec{x})\,\vec{x}\,dA + \int_{\Omega^{\text{hole+border}}} (\rho(\vec{x}) - \rho_{\text{infill}})\,\vec{x}\,dA, \qquad (2.35)
\end{aligned}$$

where Ω denotes the cross section of the square undisturbed by a hole. Under our assumption of a square with uniform density the first integral in this expression equates to the zero vector. That is, for a square without the off-center hole the center of gravity is located at $(0,0)$. This requires that the infill is sufficiently symmetric about the center of the square so that it negligibly moves the center of gravity away from $(0,0)$.[1] It follows that for the off-center square the center of

[1]This appears to be a good approximation for the *grid* infill pattern but not, for example, the *cat* infill pattern or for *grid* at very low infills.

gravity can then be estimated as

$$\begin{aligned}
\frac{M_{\text{obj}}}{L}\vec{G} &= \int_{\Omega^{\text{hole+border}}} (\rho(\vec{x}) - \rho_{\text{infill}})\, \vec{x}\, dA, \\
&= (\rho_{\text{nail}} - \rho_{\text{infill}}) \int_0^{2\pi} \int_0^{r_H} \vec{x} r\, dr\, d\theta \\
&\quad + (\rho_{\text{PLA}} - \rho_{\text{infill}}) \int_0^{2\pi} \int_{r_H}^{r_H+t} \vec{x} r\, dr\, d\theta,
\end{aligned} \tag{2.36}$$

where each of the density terms in these expressions are assumed to be independent of position. These integrals can be evaluated writing $\vec{x} = (a + r\cos\theta, b + r\sin\theta)$. It follows that

$$\begin{aligned}
\frac{M_{\text{obj}}}{L}\vec{G} &= \pi(\rho_{\text{nail}} - \rho_{\text{infill}}) r_H^2 (a,b) + \pi(\rho_{\text{PLA}} - \rho_{\text{infill}})[(r_H + t)^2 - r_H^2](a,b), \\
&= \left\{\pi(\rho_{\text{nail}} - \rho_{\text{infill}}) r_H^2 + \pi(\rho_{\text{PLA}} - \rho_{\text{infill}})[2r_H t + t^2]\right\}(a,b).
\end{aligned} \tag{2.37}$$

So, for the off-center square with hole at (a, b) the center of gravity is shifted towards (a, b) from the origin by terms proportional to density differences and cross-sectional areas. Note that all of the quantities in this expression can be determined by straightforward measurements and are listed in Table 1.

Practically speaking, our prints are not completely uniform in the direction orthogonal to the square face since the top and bottom square faces are solid PLA. An improved estimate for the center of gravity that accounts for the two end faces of the square of thickness t with density ρ_{PLA} is

$$\begin{aligned}
M_{\text{obj}}\vec{G} &= \Big\{ \left[(\rho_{\text{nail}} - \rho_{\text{infill}})(\pi r_H^2) + (\rho_{\text{PLA}} - \rho_{\text{infill}})\pi(2r_H t + t^2)\right](L - 2t) \\
&\quad + 2t(\rho_{\text{nail}} - \rho_{\text{PLA}})(\pi r_H^2)\Big\}(a,b), \\
&= M_{\text{nail}}(a,b) + \Big\{ \left[-\rho_{\text{infill}}(\pi r_H^2) + (\rho_{\text{PLA}} - \rho_{\text{infill}})\pi(2r_H t + t^2)\right](L - 2t) \\
&\quad - 2t\rho_{\text{PLA}}(\pi r_H^2)\Big\}(a,b).
\end{aligned} \tag{2.38}$$

That is, the new center of gravity, $\vec{G}$, is shifted towards the hole location (a, b) by an amount related to the mass of the nail (we use $M_{\text{nail}} = 0$ for an open hole) and terms related to the thickness of the hole and the material it replaces (either infill or boundary).

The predicted effective density for the block of square cross section is

$$\rho^{\text{eff}}_{\text{square}} = \frac{M_{\text{square}}}{V_{\text{ext}}}, \tag{2.39}$$

where the total mass of the object is

$$M_{\text{square}} = (V_{\text{ext}} - V_{\text{int}})\rho_{\text{PLA}} + V_{\text{int}} I \rho_{\text{PLA}}, \tag{2.40}$$

and $V_{\text{ext}} = s^2 L$ and $V_{\text{int}} = (s - 2t)^2(L - 2t)$. When the block is printed with a hole of radius r_H parallel to the long axis the predicted effective density is

$$\begin{aligned}
\rho^{\text{eff}}_{\text{square + hole}} &= \frac{1}{V_{\text{ext}}}\Big\{M_{\text{square}} + M_{\text{nail}} - \pi r_H^2[2t + (L - 2t)I]\rho_{\text{PLA}} \\
&\quad + \pi[(r_H + t)^2 - r_H^2](L - 2t)(\rho_{\text{PLA}} - I\rho_{\text{PLA}})\Big\}
\end{aligned} \tag{2.41}$$

TABLE 1. Various parameter values for 3D prints with square cross section and a hole. The values of M_{nail} and ρ_{nail} were obtained by noting that each nail was 60 mm in length and 2 mm in radius and that 25 nails weighed 137.51 g.

Parameter	Description	Value
I	Infill Fraction (Infill % /100)	$0.05 \ldots 0.95$
s	Length of Side of Square	30 mm
L	Length of Object	60 mm
ρ_{PLA}	Density of PLA	1.15 g cm^{-3}
M_{nail}	Mass of Nail	5.5004 g
ρ_{nail}	Density of Nail $M_{\text{nail}}/V_{\text{nail}}$	7.295 g cm^{-3}
ρ_{infill}	Effective Density of Infill $\rho_{\text{PLA}} \times I$	varies with Infill
r_H	Radius of Hole	2.5 mm
t	Thickness of Solid Border	0.8 mm

The effective density of the object without the nail filling the hole is calculated by the same formula with M_{nail} set to zero.

Various comparisons between this theory and our experimental observations and measurements are given below. First, however, we need to create our floating objects.

3. Methods for 3D printing

In order to experiment with floating objects, we have opted to experiment by designing and 3D printing them. This has the advantage that we can easily create any object we can describe mathematically. In addition, we can vary the density of our print by changing the infill density, a parameter which is set at the time of printing. In this section, we describe the full workflow needed – and we have kept our presentation accessible to those with no 3D printing experience, in the hopes of making it possible for anyone with an interest to create their own experiments. The process consists of three steps: first, we need to design the objects in design software. We then need to give the print specific parameters, within a *slicer* software, where the choice of slicer software depends on the printer. Finally, we print the objects on a 3D printer. We detail the workflow of these steps here.

3.1. Design. We have opted to design the objects in OpenSCAD [**OS**], which is a free command line computer aided design (CAD) system. Figure 3 shows three of our experimental floating objects generated in OpenSCAD. Since we are interested in creating objects with a fixed cross section, we are able to do so in just a few lines of code. For convenience, we include syntax here so a reader could create their own. Copies of both sample code and stl files are available from [**GIT**].

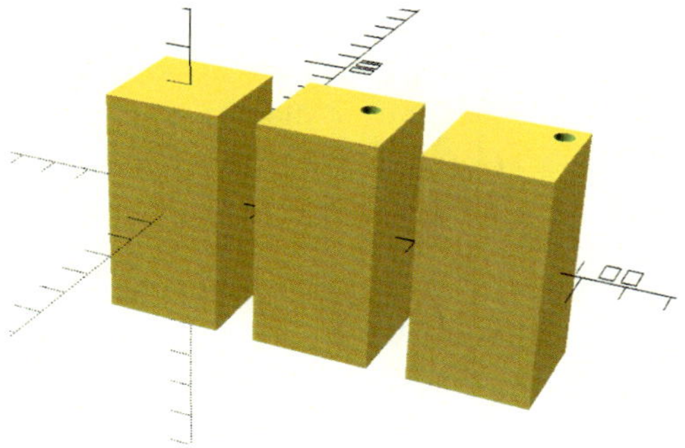

FIGURE 3. This shows three cubes in OpenSCAD. The leftmost cube is at the "true" center. The other two cubes have a hole on the diagonal line of the cross section. All cubes are shown with the longest direction vertical. This is the orientation in which we printed all cubes.

To create a box with height 60 mm and a 30 mm by 30 mm square cross section, we use the command

```
cube([30,30,60]);
```

For our off-center weight experiments, we have placed a hole of radius 2.5 mm lengthwise in the interior of the box. This is done by taking the set difference between the box above and a cylinder. In order to create a cylinder of height 60 mm and radius 2.5 mm, we use the command

```
cylinder (h = 60, r=2.5, center = true, $fn=100);
```

The command `center = true` centers the object so it is easier to position it with respect to the cube. The command \$fn=100 indicates that the circle should be estimated by a 100-sided polygon. In order to have the cylindrical hole to be displaced from the center of the cube, we use the `translate` command. Putting this all together to create a vertical hole that is displaced by 10 mm diagonally from the center of the cube, we use the command sequence:

```
difference() {
   cube([30,30,60], center =true);
   translate([10,10,0])
     cylinder (h = 80, r=2.5, center = true, $fn=100);
}
```

While most of our floating objects were simple cubes with and without holes, in Section 5.2 we also considered more complicated objects, where the cross section was given as a polygon in the form $\{(x_1, y_1), (x_2, y_2), \ldots, (x_n, y_n)\}$. This includes for example the Mason M in Figs. 1 & 15. In order to create objects with general polygonal cross sections in OpenSCAD, we first create a polygon with the `polygon` command, and then we can turn this into a solid with fixed cross section using the `linear_extrude` command. For example, the following creates a cylinder of height 60mm with a cross section which is a polygon determined by the ordered points $(0, 0), (50, 0), (60, 40), (50, 20)$.

```
points=[[0,0],[50,0],[60,40],[50,20]];
linear_extrude(height=60) polygon(points);
```

3.2. Slicing and printing. We now describe the printing parameters set within the slicing software. In order to speed up printing and keep objects light, a 3D printed object is usually partially hollow inside, printed with a lattice pattern filling a fixed *infill fraction* of the object interior. We denote this quantity by I. The fact that we can vary the infill fraction is quite useful for us, since our goal is to see how the stable floating orientations change with density. To be consistent, we did not vary the *infill pattern*; we printed all of our prints with the *grid* infill pattern. For further consistency, all of our floating objects were printed on a Makerbot Replicator 5th Generation printer and sliced in Makerbot Print proprietary software.

The mass of a print is most significantly affected by the infill fraction, but this is not the only factor. In addition, an outer border layer of the print is printed at 100% infill for a fixed thickness on the sides, and a fixed thickness on the top and bottom. For the most part, the default values for all thickness is 0.8 mm, the thickness of two *shells*, where a shell is 0.4 mm, i.e. the diameter of a standard extruder nozzle.

In order to calculate the predicted mass of a print, we need to know ρ_{PLA}, the density of PLA. Most sources state that this value is around 1.24 – 1.25 g cm^{-3}, though [**DAetal**] used the value 1.17 g cm^{-3}. This number can vary quite significantly depending on the brand of PLA, and therefore we decided to measure the value for our lab conditions. In particular, we printed a number of squares that were 30 mm by 30 mm by 1.6 mm. Since the the top and bottom of a print are 0.8 mm solid filament, the print is guaranteed to be 100% infill. We printed fifteen test tiles, roughly half with Makerbot brand filament and half with an off brand filament. Most of our test tiles were printed in part on the same Makerbot printer and slicer as the floating objects, but we also printed some on a Monoprice Mini. We found that the printer used was not an important factor for the mass of the test tiles, but the brand was an important factor. For Makerbot brand filament and off-brand filament respectively, our measured densities were 1.082 ± 0.017 and 1.152 ± 0.010 in g cm^{-3}. We have chosen the value $\rho_{PLA} = 1.15$ g cm^{-3} for our calculations.

Taking into account the infill percentage, shells, and measured value for ρ_{PLA}, we expect the mass of our object to be given by ρ_{PLA} times V_{ext}, the volume of the outer layer plus ρ_{PLA} times I, the infill fraction, times V_{int}, the volume of the interior of the print. If there is a nail embedded in the hole, then we additionally have to include this value in our calculation. We have described this calculation above in (2.41). Figure 4 shows the predicted values for prints without a nail hole (dashed red) with a nail hole (solid green). We also show the measured values for the prints we have made without a nail hole (red x) and with a nail hole (green o). Note that while quite useful to be able to predict mass of a print from the infill density, all of our calculations of R that we used for particular prints in our experiments are based on the mass of the print measured directly using a digital scale.

It is critical for the print to be uniform in the longest direction, as our calculations consider this direction as completely invariant. Therefore we printed with the long direction going from top to bottom so that the infill is identical at each cross section.

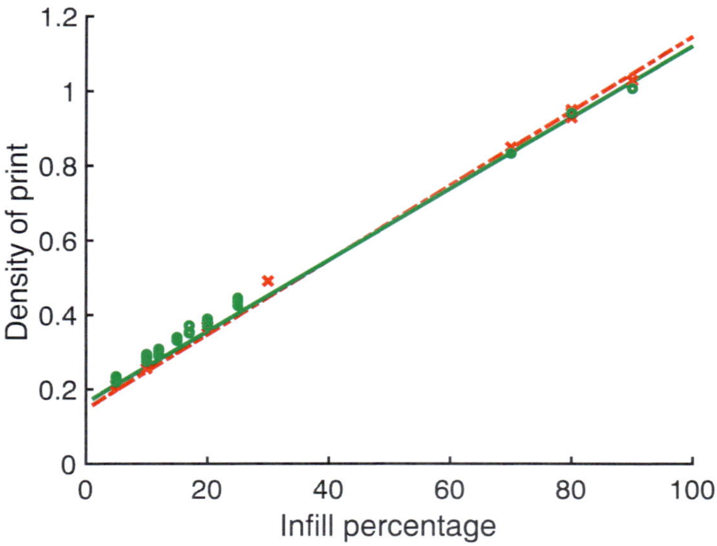

FIGURE 4. Density of prints as a function of infill. Prints have 30mm by 30mm cross section and 60mm length. The predicted values without hole (dashed red) and with hole (solid green), graphed along with the measured values without hole (red x) and with hole (green o).

Since our prints were printed on a Makerbot printer, we have used the Makerbot proprietary slicing software Makerbot Print. However, there is nothing about the process that could not be modified to other slicer software.

The objects are ready to float!

4. Experiments and data acquisition

Once the objects were printed our objectives turned to floating these objects and obtaining measurements associated with the stable floating orientations that we could compare with the theory. After floating the objects in a test tank to observe qualitative behavior (e.g. confirming the objects floated, identifying the number of stable orientations, etc.) we turned to gathering quantitative information in the form of stable floating orientations.

We used a tank with clear flat sides and filled with tap water (see Figure 5). In order to measure angles associated with a stable floating orientation the object was carefully placed by itself in the tank and allowed to come to rest. In order to ease the task of keeping the object in a position with its square cross section facing the side of the tank where the camera was positioned we placed vertical guides a small distance away from the object. One of these guides was a ruler that we could later use in the image analysis for calibration. A typical view of a floating object is shown in Figure 6.

Digital images were obtained with a Canon EOS Rebel XSi used in manual focus mode. A remote shutter release was used to avoid bumping the camera, which was mounted on a tripod sitting on the same table as the tank. Care was also taken to line up the camera at the water level and to have the object floating

FIGURE 5. Our experimental setup involves floating 3D printed objects mostly with square cross sections. An occasional print had density ratio $R > 1$ and did not float.

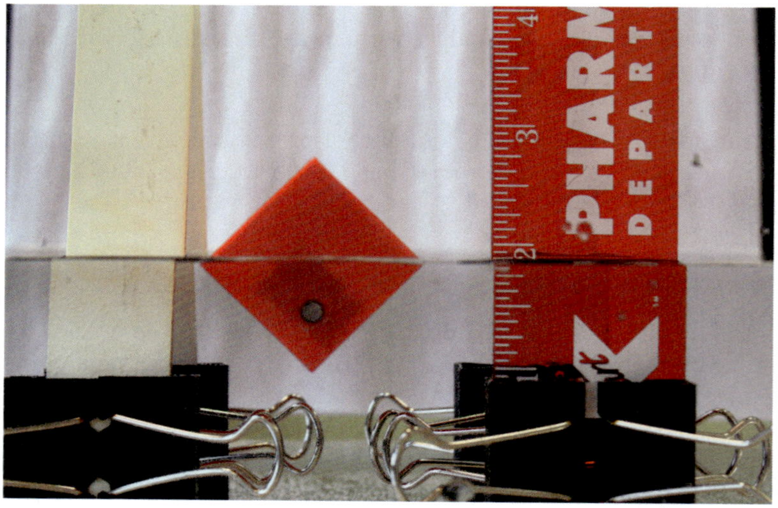

FIGURE 6. A typical view of the experimental set up to measure angles associated with stable floating orientations. Vertical guides were used as an aid to keep the object floating in the part of the tank on which the camera was focused. Care was taken to assure that contact with underwater objects was avoided and any incidental contact of the object with the guides did not alter its floating orientation. We also took care to prevent contact of object with the sides of the tank.

in a direct line to the camera. A certain amount of refraction could be observed, especially in directions off the main viewing axis.

We used Matlab's grabit.m software to extract information from each image such as the one in Figure 6. We first calibrated the view by selecting points on the ruler in view on the right, taking care to use points above the waterline to avoid distortion of the ruler by the water. For a floating square we then identified 6 points in the image. Two of these points were chosen on the waterline relatively far from the object on both sides so that the orientation of the waterline could be obtained. Four other points were chosen at the four corners of the square (clockwise starting from the top of the square). With this information we obtained four vectors by taking differences of adjacent vertices and used $\cos\theta = a^T b/\|a\|_2\|b\|_2$ where b is a waterline vector and a is one of square-side vectors. In the example of Figure 6 one pair of sides (northwest and southeast) were used to identify an estimate for θ while the other pair of sides (northeast and southwest) were used to estimate the complementary angle with respect to $\pi/2$. In the case of the Mason M, we used a similar approach but chose four points along the bottom two 'legs' of the M to make angle measurements.

For prints with holes, we covered the holes (void or filled with a nail) with waterproof tape to keep water out. In cases with no hole or when the hole was in the center of the object, we used a very small amount of nail polish to mark one corner of the square. This allowed us to assess asymmetry of the object that was present either unplanned or by design.

For each object we first measured its mass (with hole either left as a void or filled with a nail) using a digital scale and from that obtained an effective density according to $\rho_{\text{obj}} = M_{\text{obj}}/V_{\text{obj}}$. This corresponds to the material density for a uniform object or the effective uniform density for a non-uniform object. The volume of the object V_{obj} was computed either from measurements of the dimension and known formulas (e.g. width times length times height for a rectangular box) or using polyarea in Matlab (for more general cross sections) multiplied by the 3D print scaling.

5. Computational approaches

Various Matlab codes were developed to compute results and analyze our floating shapes.

5.1. Square cross sections. Various Matlab codes developed for the square cross sections have been posted in a GitHub repository [**GIT**]. These include

- SQUARE_PE_GxGy.m: This code is based on the potential energy formulas outlined in the section on the square. It generates for given values for the density ratio R and the center of gravity (G_x, G_y) the computed potential energy landscape and includes options to plot the potential energy landscape and the square floating in a stable orientation. This code was used to generate the theoretical predictions in Figures 8, 9, and 10, for example.
- SQUARE_ANGLES_GxGy_R_Looper.m: This code is based on the formulas given in the section on the square and plots for a given center of gravity (G_x, G_y) and specified range $R \in [R_{min}, R_{max}]$ the stable orientation angles. This code, for example, was used to generate Figures 7 and 13.

5.2. General polygonal cross sections. In the case of a long floating object of uniform density with a general polygonal cross section, though it is no longer possible to give as detailed an analysis as in the case of a square, we are still able to apply Archimedes' Principle and calculate the center of gravity, center of buoyancy, and the potential energy of a floating configuration, as we describe below.

5.2.1. *Computation of stable floating configurations.* When doing the calculations for a general polygon, we wrote a program that takes two vectors with the x and y values of our polygon and a density ratio and goes through the following algorithm. Firstly, it takes the shape of the given polygon and calculates the center of gravity of the object, assuming a uniform density throughout. Secondly, we identify the correct placement of the waterline determined by Archimedes' Principle that establishes the correct submerged area to total area ratio. This is done by use of a bisection method where at an orientation we take the lowest point of our object and create a waterline through it and make that our lower bound. We then take the highest point of our object and make that our upper bound. We then calculate the area ratio of each of our bounds and find the midpoint of our upper and lower bounds and calculate the area ratio for the waterline going through the midpoint. If the waterline through the midpoint is above the correct placement, it becomes our new upper bound and if it is below the correct placement, then it becomes the new lower bound. We apply the bisection method until the correct waterline is found for our original orientation. The correct placement of our waterline in the case of uniform density is where the waterline splits the object into two areas where the submerged area relative to the total area is equal to the desired density ratio (e.g. for an iceberg with uniform density, the line will be such that the new submerged area created by the line relative to the area of the original polygon will match the density ratio 0.8912). Thirdly, we compute the center of buoyancy of the object by applying the same method used for the center of gravity except using the submerged area determined by the polygon and the waterline. Finally, we calculate the potential energy function for our polygon at the orientation and repeat the process for all angles.

The input to this code is a planar polygonal region, oriented counterclockwise and a density ratio of the object relative to the water. The output is a plot of the potential energy landscape with respect to the angle. By default, the computation is performed for uniform density. However, it is possible to compute this information for objects with non-uniform density if one inputs the center of gravity.

6. Results

6.1. Floating squares: symmetric case. Figure 7 shows stable equilibrium angles as a function of density ratio R for objects with square cross sections and center of gravity at the center of the square, $\vec{G} = (0, 0)$. Various experimental results are shown for 3D printed shapes with different effective densities. These effective densities have been modified as described earlier by adjusting the infill as well as printing objects with a hole at the center which we can leave as void space or fill with a denser object, such as a nail.

As a visualization, we show several potential energy landscapes and selected shape orientations predicted from the theory and observed experimentally. Figure 8 shows the potential energy landscape for $\vec{G} = (0, 0)$ for a case with $R = 0.23296$ which corresponds to a region in parameter space where eight stable orientations exist. The eight orientations in this case come in pairs, as indicated in the lower portions of Figure 8. The eight experimentally-observed orientations are shown

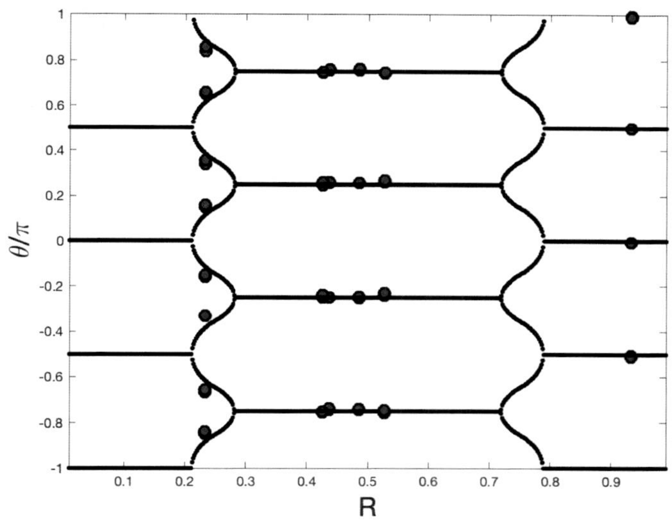

FIGURE 7. This plot shows stable floating orientations versus density ratio R for $\vec{G} = (0,0)$. The marks show various measured equilibrium orientations for several of our 3D printed objects.

by the points at $R = 0.23296$ in Figure 7.[2] Figure 9 shows the potential energy landscape for $\vec{G} = (0,0)$ for a case with $R = 0.4856$ which corresponds to a region in parameter space where four stable orientations exist. These orientations correspond to the object floating with the corner straight up. Experimental measurements for these angles correspond the angle measurements shown at $R = 0.4856$ in Figure 7. Figure 10 shows the potential energy landscape for $\vec{G} = (0,0)$ for a case with $R = 0.9322$ which corresponds to a region in parameter space where four stable orientations exist. These orientations correspond to the object floating with the flat side of the square straight up. Experimental measurements for these angles correspond the angle measurements shown at $R = 0.9322$ in Figure 7.

A more thorough experimental exploration of the parameter space, particularly in the region around $R = 0.25$, where eight stable orientations can be identified has recently been done by Feigel & Fuzailov [**FF**]. Those authors used a larger floating object (114 mm $\times$ 114 mm $\times$ 353 mm) constructed using two tin tea boxes that could be fitted/weighted with additional bars and magnets to adjust the object's effective density.

6.2. Floating squares: breaking symmetry. Figure 11 show the potential energy landscape for three different prints corresponding to square cross sections with nail-filled holes at **A**: $(0,0)$ (upper left plot), **B:** $(0.3, 0.3)$ (upper right plot), and **C**: $(0.45, 0.45)$ (lower plot). Note that the coordinates for the holes are given in units of $s/2$ where s is the length of the side of the square. These three have

[2] A keen eye will note both blue dots and red dots in this sequence in Figure 7. These two different sets of angle estimates correspond to the two different angle measurements described in the earlier section on Experiments and Data Aquisition.

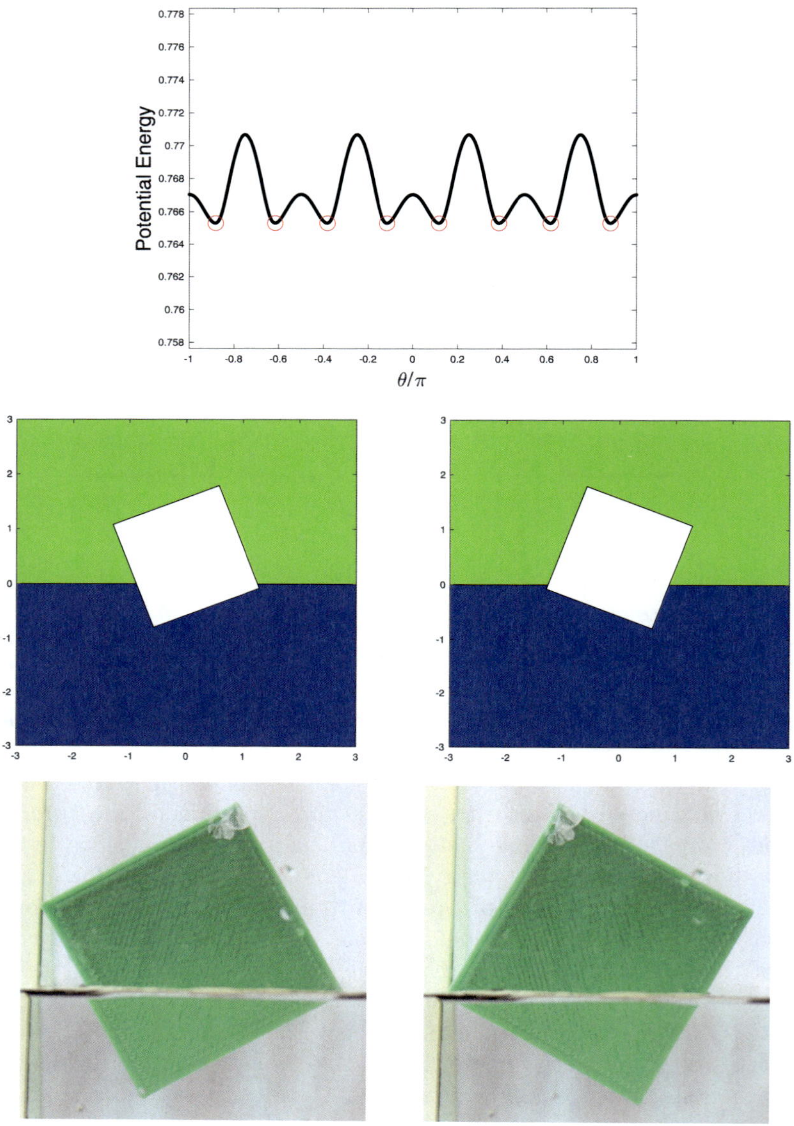

FIGURE 8. The upper plot shows the potential energy landscape for the square with $\vec{G} = (0, 0)$ and $R = 0.23296$. There are eight stable equilibria. Two stable floating configuration corresponding to the orientations closest to $\theta = 0$ are shown in the plots in the second row. By symmetry these also match with the other stable orientations on $\theta \in [-\pi, \pi]$. Corresponding experimental images are also shown in the bottom row. Measured angles for this case are shown in Figure 7.

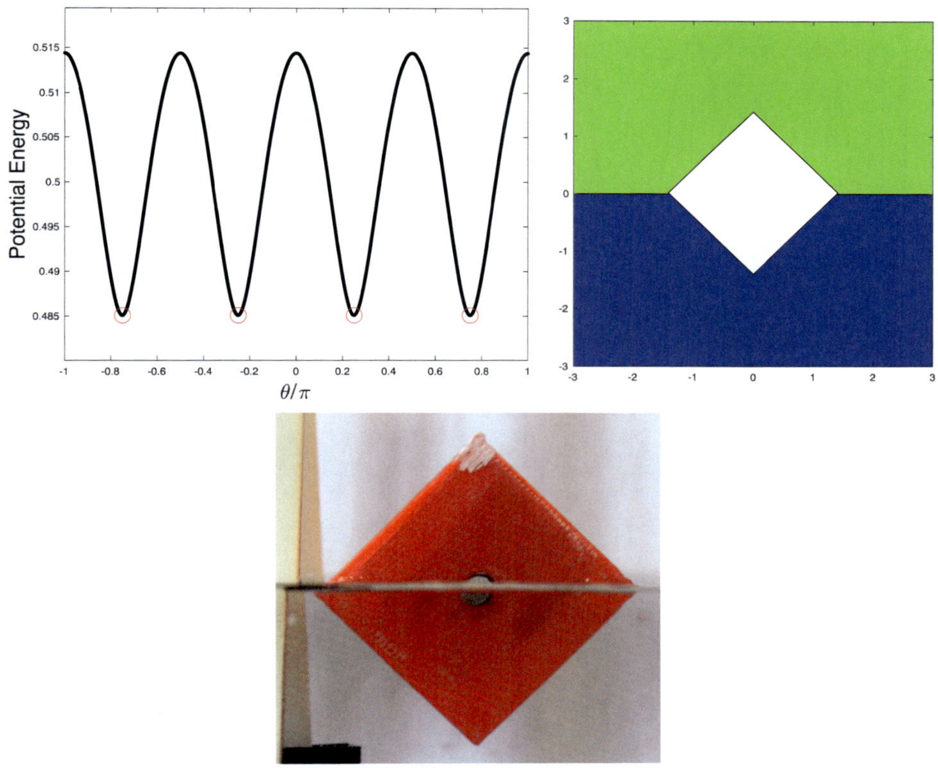

FIGURE 9. The upper left plot shows the potential energy landscape for the square with $\vec{G} = (0, 0)$ and $R = 0.4856$. There are four stable equilibria corresponding to $\theta = \frac{\pi}{4} \pm \frac{\pi}{2}n$ for integer n. The upper right plot shows that the square floats with vertex pointing upwards. These four stable orientations are also observed experimentally (one such orientation is shown in the image). Measured angles for this case are shown in Figure 7.

density ratios of $R = 0.4856$ for Case **A**, $R = 0.4874$ for Case **B**, and $R = 0.4911$ for Case **C**.

For Case **A** with nail-filled hole at $(0, 0)$ already discussed in Figures 7 and 9, four stable orientations, with a corner of the square pointing straight up, are predicted theoretically and observed experimentally.

Experiments for Case **B**, with a nail-filled hole at $(0.3, 0.3)$, are shown in Figure 12. In this case the center of gravity $\vec{G}$ is no longer at the center of the square and the symmetry is broken. Despite this broken symmetry, four stable orientations are still observed experimentally as shown in Figure 12. Our theory predicts that $(G_x, G_y) = (0.06789, 0.06789)$ and the corresponding potential energy plot is shown by the solid curve in the upper right plot of Figure 11. There are only two stable orientations predicted at this value of $\vec{G}$ and so our theory does not match the experimental observations. However, two other curves are shown in the upper

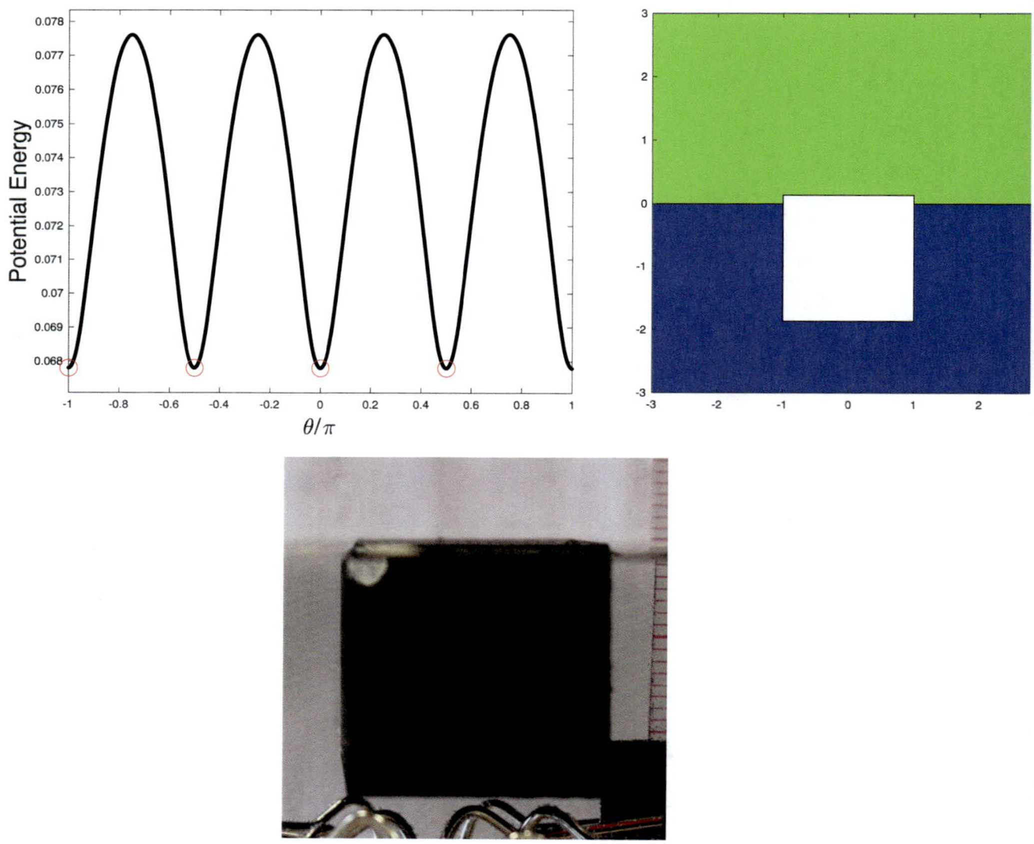

FIGURE 10. The upper left plot shows the potential energy landscape for the square with $\vec{G} = (0, 0)$ and $R = 0.9322$. There are four stable equilibria corresponding to $\theta = 0 \pm \frac{\pi}{2} n$ for integer n. The upper right plot shows that the square floats deep in the water with flat side up. These four stable orientations are also observed experimentally (one such orientation is shown in the image). Measured angles for this case are shown in Figure 7.

right plot in Figure 11 – the dashed curve has $(G_x, G_y) = (0.05, 0.05)$ and the dash-dotted curve has $(G_x, G_y) = (0.04, 0.04)$. These curves with nearby values of $\vec{G}$ indicate that there are two other stable orientations nearby. One explanation for the discrepancy between experiment and theory is that the center of gravity of our print is not exactly where we predict it to be, perhaps due to uncertainties in the infill structure of the print. Another potential source of imprecision in the center of gravity is the nail not fitting precisely in the middle of the hole, but the nail seems to fit snugly in the hole, so it seems a less likely explanation. Also, there is evidence of menisci at the solid–liquid–air contact line in Figure 12 suggesting that surface tension could provide a large enough force to hold the print in an otherwise slightly unstable configuration.

For Case **C**, with a nail-filled hole at (0.45, 0.45) and $(G_x, G_y) = (0.1017, 0.1017)$ the potential energy landscape shown in the lower plot of Figure 11 indicates that only two stable equilibria exist. For this case both theory and experiment are in agreement on the number of stable equilibria. The experimental images for this case are not shown but are very similar to the upper left and lower left images of Figure 12. Corresponding stable orientations with the nail to the left or right as in the upper right and lower right images of Figure 12 no longer exist for Case **C**.

As we have just observed, when $\vec{G} \neq 0$ the symmetry of the square is broken and the predicted number of stable equilibria change. We display this in Figure 13 for four cases $\vec{G} = (0.01, 0.01)$, $(0.02, 0.02)$, $(0.04, 0.04)$, and $(0.06789, 0.06789)$. On the last two of these we have plotted the four experimentally-observed stable orientation angles for the print denoted Case **B** with a nail-filled hole at $(0.3, 0.3)$. As just discussed for Case **B**, our theory predicts $\vec{G} = (0.06789, 0.06789)$ and only two equilibrium values of θ exist for this case as shown in the lower right plot of Figure 13. Reduction in the value of $\vec{G}$ to $(0.04, 0.04)$, for example as indicated in the lower left plot of Figure 13 shows four stable orientations at the density ratio, $R = 0.4874$, of Case **B**. The upper two plots corresponding to $\vec{G} = (0.01, 0.01)$ and $(0.02, 0.02)$. Especially in comparison to the predicted stable angles shown in Figure 7 these plots reveal interesting bifurcation structure as the square symmetry is broken. Note that in addition to floating squares with four or eight stable orientations, when the symmetry is broken in this way (moving the center of gravity towards a corner) there are also situations where either three or six stable orientations are predicted.

6.3. Results for general polygonal cross sections: the mason M. As an example of a floating shape with nontrivial cross sectional area we chose the Mason M. This print was made based on a counter-clockwise-oriented set of points describing the shape of the M.

The cross-sectional area of the Mason M, A_M, was obtained using Matlab's `polyarea.m` applied to the point set described above. This value adjusted by a pixel to millimeter (mm) conversion factor used to scale for 3D printing gave the area. In particular, we found that $A_M = 126,550 \cdot (0.06)^2 \text{ mm}^2 = 455.58 \text{ mm}^2$. The length of our Mason M was 70 mm, giving a volume of $31,891 \text{ mm}^3$. Since its mass was 27.92g its corresponding effective density is $\rho_{\text{eff}} = 0.8755 \text{ g cm}^{-3}$, which, in comparison to the density of water, is fairly close to the typical density ratio of an iceberg in a polar sea. Figure 14 shows the potential energy landscape along with predicted stable configurations for the Mason M. The four stable orientations of the floating Mason M are shown in Figure 15.

Figure 16 shows another variation of the full Mason M (with feathers) inspired by calving events that can happen with real icebergs. These two plots show stable floating orientations, computed using the general polygonal code, where a *calving* event of shedding a feather results in a new stable floating orientation (right plot). The red line in the plot on the right shows the pre-calving waterline position.

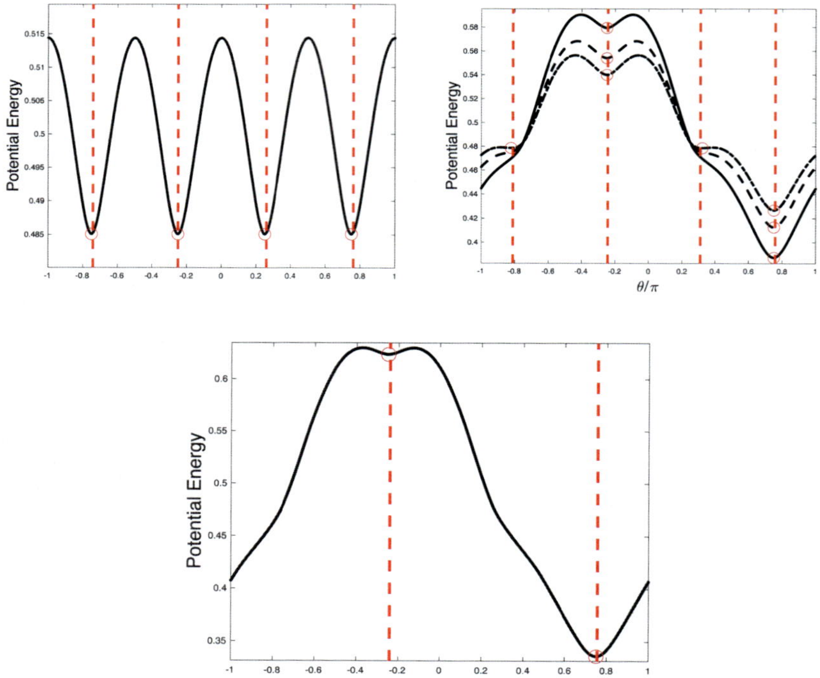

FIGURE 11. Potential energy plots for squares with nail-filled holes at **A**: $(0,0)$ (upper left plot), **B:** $(0.3, 0.3)$ (upper right plot), and **C**: $(0.45, 0.45)$ (lower plot). The black curves (solid, dashed, or dash-dotted) show the potential energy function defined in equation (2.33). The open red circles indicate the theoretical local minima of the potential energy. The vertical red dashed lines show the angles at which experimentally-floating squares appear to be stable. Case **A** has four-fold symmetry and four stable orientations are predicted theoretically and observed experimentally (see also Figure 9). For Case **B**, in which the symmetry is broken, our theory predicts only two stable orientations but we observe four experimentally (these four orientations are shown in Figure 12). The dashed and dash-dotted black curves in the upper right plot show two other potential energy landscapes for nearby values of $\vec{G}$ (see text for details) indicating the presence of nearby stable states. For Case **C**, the square is farther from symmetric and both theory and experiment predict only two stable orientations. We do not show experimental images for Case **C**, but they are similar to the upper left and lower left images in Figure 12.

7. Conclusion and open problems

7.1. Motivation. The equilibria of floating bodies have been studied since antiquity. Results on their stability associated with metacentric concepts that date

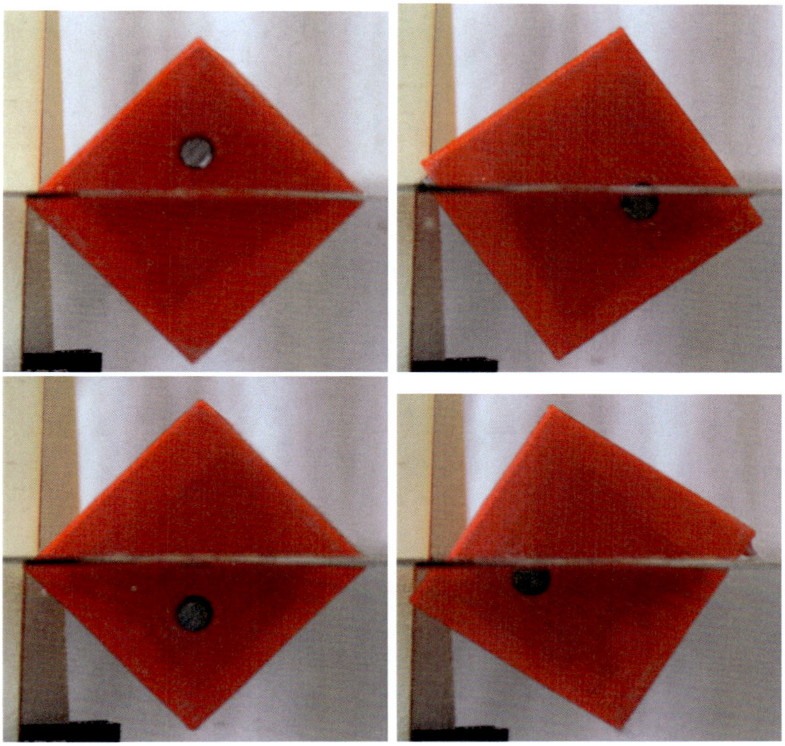

FIGURE 12. Four floating orientations for Case **B** in Figure 11 with a nail-filled hole at $(0.3, 0.3)$ giving a nonzero center of gravity $\vec{G}$. When the hole is *up* or *down* the stable orientations correspond to corner straight up. When the hole is *left* or *right* the top corner of the square is deflected counterclockwise or clockwise, respectively.

back to eighteenth-century continue to play a major role in areas of naval architecture and the study of icebergs.

The field has recently gained mathematical interest, especially on Ulam's floating body problem. In problem 19 from the Scottish Book Ulam asks: "Is a solid of uniform density which will float in water in every position a sphere?" [**M**]. Counterexamples to Ulam's problem in the plane for $\rho = 1/2$ go back to Auerbach [**Au**], and for $\rho \neq 1/2$ to Wegner who also obtained results for non-convex bodies (holes in the body are allowed) in [**W2**]. Most recently Florentin *et al.* [**F**] gave an affirmative answer for a class of origin symmetric n-dimensional convex bodies with $\rho = 1/2$ relative to water. While counterexamples in the plane exist, Ulam's floating body problem in higher dimensions is open to the best of our knowledge.

7.2. Limitations. Using 3D printed models we were able to produce and collect experimental results in agreement with the theory. Our results and experimentation process can be reproduced readily, increasing access to 3D printers within educational facilities gives way for students to experiment with stability and generate different scenarios including Ulam's floating body problem.

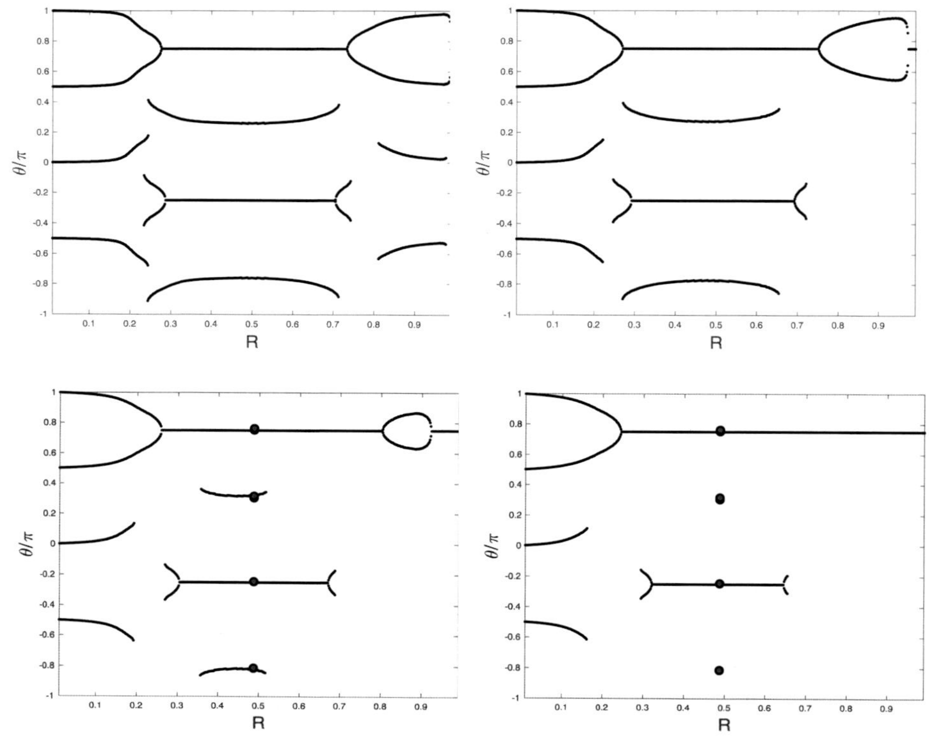

FIGURE 13. Stable equilibrium angles versus density ratio R for off-center squares. These plots have $\vec{G} = (0.01, 0.01)$ (upper left), $(0.02, 0.02)$ (upper right), $(0.04, 0.04)$ (lower left), and $(0.06789, 0.06789)$ (lower right). Experimentally-measured equilibrium angles for Case **B** corresponding to $R = 0.4874$, described in the text and in Figures 11 and 12, are shown in both the lower left and lower right plots. The full sequence of plots in this figure should be compared to the one for the symmetric square in Figure 7 which has $\vec{G} = (0, 0)$.

The biggest challenge faced when producing our 3D samples is related to density. To design infill patterns and infill densities using computer-aided design(CAD), we produced samples to measure precision and increase the accuracy of our experiment and identify possible sources of experimental errors. A limitation in the 3D printing process is the difficulty in producing low-density objects to desired accuracy. To maintain an even distribution of mass, considerations must be taken with respect to the infill pattern. Zigzag, grid, triangle, and concentric patterns are recommended when an even distribution is desired. While Makerbot's cat infill model is not recommended for experimenting with homogenous objects as it is asymmetric, it made for interesting observations and ultimately motivated further investigation into non-homogeneous objects, and it's also really cute, cf. Figure 17.

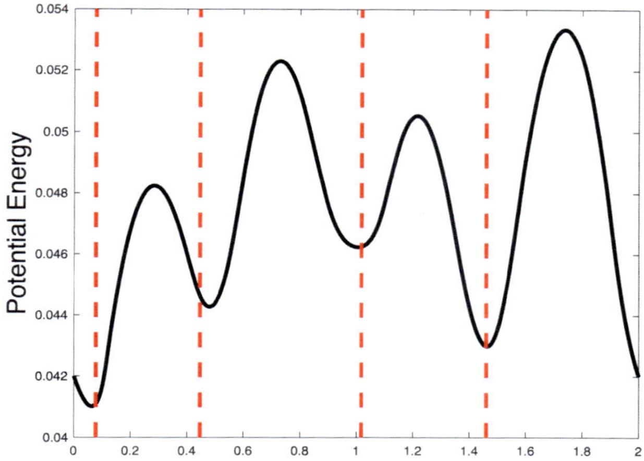

FIGURE 14. The Mason M potential energy plot where potential energy depends on the orientation angle of the waterline with respect to the 3D Mason M Print (zero angle corresponds to an M in its usual upright orientation). Each vertical red dotted line corresponds to an experimentally found stable orientation. The local minima of the graph define stable orientations theoretically calculated using our code. See also Figure 15.

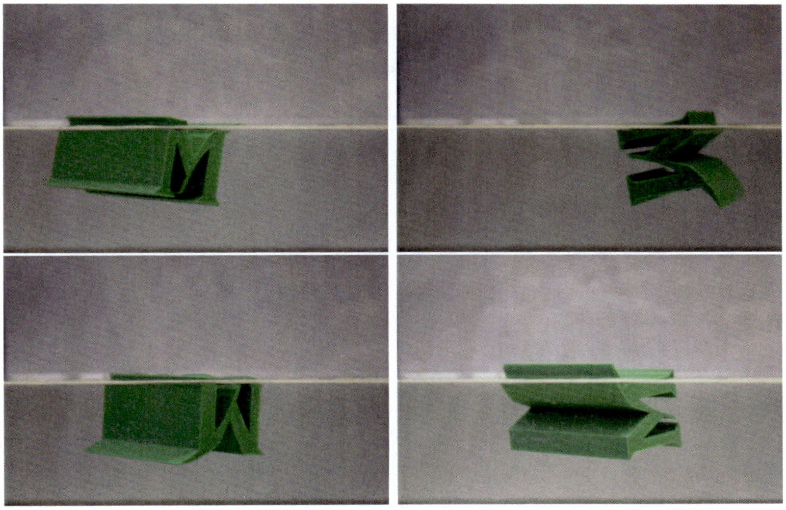

FIGURE 15. The four stable orientations of the floating Mason M. The top left orientation matches up with the left most red dotted line in Figure 14. The top right orientation matches up with the second left-most red-dotted line in Figure 14. The bottom left orientation matches up with the second-right most red dotted line in Figure 14. The bottom right orientation matches up with the right-most red dotted line in Figure 14.

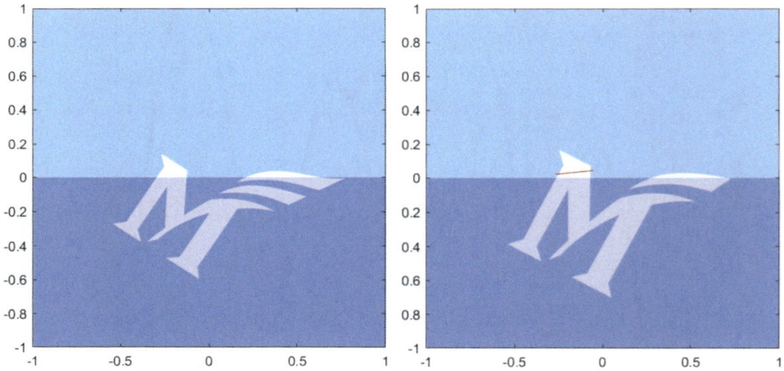

FIGURE 16. The full Mason M (with feathers) and its floating orientation after a *calving* event where the upper feather is removed. The red line shows pre-calving waterline.

FIGURE 17. The Makerbot cat infill.

Our work here has focused on effectively two-dimensional shapes. Mathematical challenges for 3D floating shapes have been examined (e.g. Erdös *et al.* [**E2**] and Wegner [**W1**]) and of course most realistic floating objects such as icebergs are three dimensional. Approaches using 3D Print design are likely to prove highly useful for future studies in these directions.

Acknowledgments

The authors would like to thank the referee for the helpful comments that allowed us to improve this paper. We would also like to thank the Mason Experimental Geometry Lab (MEGL) and Mason's Math Maker Lab, particularly Maker Lab staff Patrick Bishop and MEGL member Will Howard, for supporting this project. We would also like to thank the Department of Physics at George Mason University for the use of a digital scale. The research of E.S. was partially supported by the Simons Foundation under Awards 636383.

References

[Al] P. E. Allaire, *Stability of simply shaped icebergs*, Journal of Canadian Petroleum Technology **11** (1972) 21–25.

[Au] H. Auerbach, *Sur un problème de M. Ulam concernant l'équilibre des corps flottants*, Studia Mathematica **7** (1938) 121-142.

[B] R. C. Bailey, *Implications of iceberg dynamics for iceberg stability estimation*, Cold Regions Science and Technology **22** (1994) 197–203.

[D] R. Delbourgo, *The floating plank*, Am. J. Phys. **55** (1987) 799–802.

[DH] M. V. Deriabyn and P. G. Hjorth, *Tip of the iceberg*, European J. Appl. Math. **20** (2009), no. 3, 289–301, DOI 10.1017/S0956792509007876. MR2511277

[E1] P. Erdös, G. Schibler, and R. C. Herndon, *Floating equilibrium of symmetrical objects and the breaking of symmetry. Part 1: Prisms*, Am. J. Phys. **60** (1992) 335–345.

[E2] P. Erdös, G. Schibler, and R. C. Herndon, *Floating equilibrium of symmetrical objects and the breaking of symmetry. Part 2: The cube, the octahedron, and the tetrahedron*, Am. J. Phys. **60** (1992) 345–356.

[FF] Y. Feigel and N. Fuzailov, *Floating of a long square bar: experiment vs. theory*, Eur. J. Phys. **42** (2021) 035011.

[F] D. I. Florentin, C. Schütt, E. M. Werner, and N. Zhang, *Convex floating bodies of equilibrium*, Proc. Amer. Math. Soc. **150** (2022), no. 7, 3037–3048, DOI 10.1090/proc/15697. MR4428887

[G] E. N. Gilbert, *How things float*, Amer. Math. Monthly **98** (1991), no. 3, 201–216, DOI 10.2307/2325023. MR1093951

[GIT] D. M. Anderson, B. G. Barreto-Rosa, J. D. Calvano, L. Nsair, and E. Sander, Supplementary Materials: Online code and files, `github.com/danielmanderson/IcebergProject`.

[Ice1] J. Tauberer, *Iceberger*, `https://joshdata.me/iceberger.html`.

[Ice2] chris@engaging-data.com, *Iceberger Remixed*, `https://engaging-data.com/iceberger-remixed/`.

[DAetal] M. Lalegani Dezaki, M. K. A. M. Ariffin, A. Serjouei, A. Zolfagharian, S. Hatami, and M. Bodaghi, *Influence of Infill Patterns Generated by CAD and FDM 3D Printer on Surface Roughness and Tensile Strength Properties*, Applied Sciences **11:16** (2021) 7272. doi: 10.3390/app11167272

[M] R. Daniel Mauldin (ed.), *The Scottish Book: Mathematics from the Scottish Café with selected problems from the new Scottish Book; Including selected papers presented at the Scottish Book Conference held at North Texas University, Denton, TX, May 1979*, 2nd ed., Birkhäuser/Springer, Cham, 2015, DOI 10.1007/978-3-319-22897-6. MR3242261

[MK] J. Mégel and J. Kliava, *Metacenter and ship stability*, American Journal of Physics **78** (2010) 738–747. doi: 10.1119/1.3285975

[OS] *OpenSCAD*, `https://openscad.org`.

[Re] W. P. Reid, *Floating of a long square bar*, Am. J. Phys. **31** (1963) 565–568.

[Ro] Chris Rorres, *Completing Book II of Archimedes's On floating bodies*, Math. Intelligencer **26** (2004), no. 3, 32–42, DOI 10.1007/BF02986750. MR2088013

[W1] F. Wegner, *Floating bodies of equilibrium in three dimensions. The central symmetric case*, `arXiv:0803.1043v2` [physics.class-ph] (2009).

[W2] F. Wegner, *From elastica to floating bodies of equilibrium*, `arXiv:1909.12596v4` [physics.class-ph] (2020).

[WD] V. Wilczynski and W. J. Diehl, *An alternative approach to determine a vessel's center of gravity: the center of buoyancy method*, Ocean Engng. **22** (1995) 563–570.

Department of Mathematical Sciences, George Mason University, Fairfax, Virginia 22030

Email address: danders1@gmu.edu

Department of Mathematical Sciences, George Mason University, Fairfax, Virginia 22030

Email address: bbarreto@gmu.edu

Department of Mathematical Sciences, George Mason University, Fairfax, Virginia 22030

Email address: jcalvano@gmu.edu

Department of Mathematical Sciences, George Mason University, Fairfax, Virginia 22030

Email address: lnsair@gmu.edu

Department of Mathematical Sciences, George Mason University, Fairfax, Virginia 22030

Email address: esander@gmu.edu

Proceedings of Symposia in Applied Mathematics
Volume **79**, 2023
https://doi.org/10.1090/psapm/079/00711

A 3D printed Arduino-powered interactive Barth Sextic

silviana amethyst, Samantha Maurer, and William O'Brien

ABSTRACT. 3D printing is a wonderful way to illustrate mathematics, producing tangible objects and bringing opportunities for engagement beyond digital and print media. By augmenting such objects with additional features, we can make our pieces interactive. This paper describes the techniques and tools used to make a series of 3D prints of the Barth Sextic, a projective surface of degree six, with various features for interactivity. The major electronic additions to our snap-together plastic model are rainbow LED pixels driven by a pair of student-programmed Arduino Pro Mini development boards to visualize the dodecahedral symmetry group.

1. Introduction

Physical mathematical objects reach into the viewer's hand, and give a tangible experience. 3D printing is a method for visualizing and adding a new kind of interactivity, while posters and images are still and can only give a fixed perspective. To further enhance the tangible experience of a physical model, we can add interactivity by augmenting our printed pieces with electronics. We recognize the impossibility of providing a complete blueprint and copy-paste-derivable source code for electrifying your project, so our goal is to describe our tools and techniques, such that you could adapt some of them to your own physical pieces.

The Barth Sextic is a projective algebraic surface with dodecahedral symmetry. This almost-smooth object is nodally singular with 65 double points, the maximum attainable by a projective surface of degree six. The nodal singularities make the surface impossible to model perfectly in physical space, so we must balance representation of exact structure with appearance and function.

We designed, fabricated, and programmed three instances of a beautiful 3D printed electronic Barth Sextic lit by Neopixels. Two larger instances use opaque white rigid plastic, and the third uses clear flexible plastic. A complete light toy consists of a fixture and a controller. The large fixtures each have 140 LEDs, with

2020 *Mathematics Subject Classification.* Primary 14-04.

Key words and phrases. Software, source code, etc. for problems pertaining to algebraic geometry, Arduino, Barth Sextic.

Support for this project came from Student Blugold Commitment Differential Tuition funds through the University of Wisconsin-Eau Claire Summer Research Experiences for Undergraduates program. This material is based upon work supported by the National Science Foundation under Grant No. DMS-1929284 while the first author was in residence at the Institute for Computational and Experimental Research in Mathematics in Providence, RI, during the Illustrating Mathematics program.

seven per chip in each cone, while the smaller fixture uses a single pixel per cone. These interactive art pieces are driven by a pair of communicating Arduino Pro Mini development boards, one in the fixture and the other in a custom controller with potentiometers, buttons, switches, and a rotary encoder. The physical model snaps together using a plug and socket at each nodal singularity, with hollow connections for interal wiring, an integrated hollow hook for hanging, and a twist-and-lock cover carrying each Neopixel chip.

So that you may replicate and build on our work and ideas, we have described the theoretical and technical aspects of our creation process, from computation to implementation. We start with a discussion of the mathematics of the Barth Sextic in Section 2. Our method for computing a digital model including the singularities is in Section 3. We will discuss the production of our 3D printed snap together model with features for electronics in Section 4. The method of communication between controller and fixture is described in Section 5. Visualization of the symmetry group of the Sextic is described in Section 6.

While the main body of this paper describes technical aspects of our art piece, our goal is to inspire our readers to pursue their own dynamic projects. Hybrid electronic 3D printed pieces are complicated to design and execute, so we present an overview of process, material, component, and mindset realizations in Section 7. And since this project was completed as part of an undergraduate research project, we conclude with a description of the many fostered skills in Section 8.

2. Mathematics of the Barth Sextic

This section provides background on the Barth Sextic as a mathematical object.

2.1. Defining equation. The Barth Sextic is a complex projective algebraic surface defined in $\mathbb{P}^3$ of degree six. The surface is defined by a single polynomial in four projective variables $[x, y, z, w]$:

$$4(\varphi^2x^2 - y^2)(\varphi^2y^2 - z^2)(\varphi^2z^2 - x^2) - (1 + 2\varphi)(x^2 + y^2 + z^2 - 1)^2w^2 = 0, \tag{2.1}$$

where $\varphi = \frac{\sqrt{5}+1}{2}$ is the golden ratio.

In this paper, we are visualizing in $\mathbb{R}^3$ and need three real variables. When we further discuss the Barth Sextic, we refer a particular affine version of the surface where we treat w as a homogenizing variable, and set $w = 1$. That is, under the affine chart $w = 1$, (2.1) reduces to

$$4(\varphi^2x^2 - y^2)(\varphi^2y^2 - z^2)(\varphi^2z^2 - x^2) - (1 + 2\varphi)(x^2 + y^2 + z^2 - 1)^2 = 0. \tag{2.2}$$

This affine version of the Barth Sextic has 20 compact pieces which we call "cones" and 20 non-compact pieces that diverge to infinity. Our model discards the non-compact pieces. Of important note, the *projective* Barth Sextic has 65 double points, the maximum attainable by a projective degree six surface [**7, 16**]. For our *real* view of the surface there are a total of 50 real finite double points: 20 at tips of the cones joining to the infinite pieces, and 30 at the points joining cones.

2.2. Symmetries of the Barth Sextic. The Barth Sextic has icosahedral/-dodecahedral symmetry. Since this paper describes collaboration with undergraduates, we felt it important to include an explicit accounting of the number of rotations of each type:

(1) **Two-fold rotational symmetry**
Two-fold rotation of the Sextic corresponds to rotating 180 degrees about an axis passing through two antipodal nodes (double points) separating two opposite pairs of cones. There are 30 such points coming in pairs, hence there are $30/2 = 15$ distinct axes. They contribute $15(2-1) = 15$ non-trivial rotations, where we have subtracted 1 to account for the trivial rotation around each axis. There are no fixed cones under non-trivial two-fold rotation.
(2) **Three-fold rotational symmetry**
Three-fold rotational symmetry can be generated by rotating 120 degrees about an axis passing through two antipodal tips of cones. There are $20/2 = 10$ such distinct axes, since they come in pairs. They contribute $10(3-1) = 20$ nontrivial rotations. In each rotation two cones are fixed.
(3) **Five-fold rotational symmetry**
The five-fold rotation comes from rotating 72 degrees about an axis passing through the center of a pentagonal face formed by the tips of five of the cones. There are $12/2 = 6$ such distinct axes, contributing $6(5-1) = 24$ nontrivial rotations. Zero cones are fixed by this rotation.
(4) **Mirror symmetry**
Cones trade places with another, across a plane of symmetry determined by the choice of any two of the two-fold axes. Four cones are fixed by each mirror symmetry.

In total, the Sextic has $15 + 20 + 24 = 59$ non-trivial rotations; adding the trivial rotation brings the total to 60. Once we take into account mirroring, there are 120 total symmetries.

3. Digitally modeling the Barth Sextic, with singularities

As a pre-cursor to 3D printing, we needed an accurate digital model of the mathematical surface, including nodal singularities. This section provides a background on various options for this step, and describes the methods we used in our project.

3.1. Previous techniques for model generation. Mathematicians have been working to visualize algebraic surfaces for quite a long time, and they have developed many techniques, from hand-sculpting plaster, to modern automated methods that we used in this paper. This section is not an exhaustive history of visualization of algebraic surfaces; instead, we aim to provide some historical and mathematical context.

3.1.1. *Technique: raytracing.* We would like to first mention the fabulous software techniques developed for producing images of implicit surfaces, particularly Surfer [**5**]. Surfer and other raytracers are also used in the 2016 AMS Visual Insight Blog post [**6**], with a number of visualizations from mathematical artists. In Figure 1, we present an image of the Barth Sextic captured using Surfer. The raytracing technique produces spectacular images, with surfaces appearing in mere seconds. Further, Surfer allows for parameterized surfaces, meaning that model parameters can be controlled in real time using sliders. Yet, because the result is an image, raytracing cannot assist in our goal of producing a physical version of the surface.

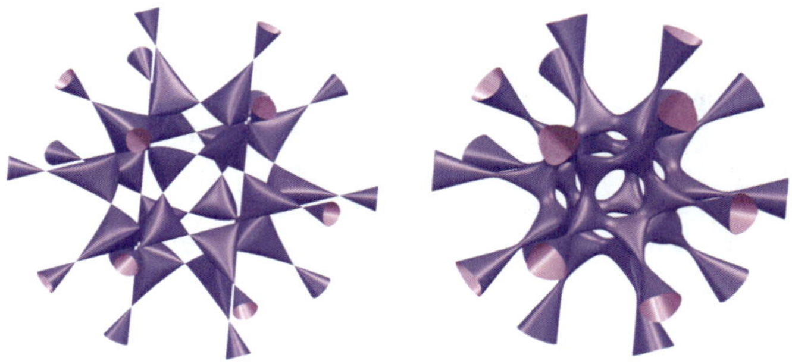

FIGURE 1. Ray traced images generated using Surfer. Beautiful, with nice reflections, but incapable of being 3D printed. On the left: the Sextic itself. Note the outer infinite pieces, and how they do not all have pixels connecting them to the inner cones. On the right: a smoothed version, produced by adding a small real number to the defining equation.

3.1.2. *Technique: smoothing.* In the physical realm, Oliver Labs has used techniques for producing smooth versions of physical models, to avoid the singularity problems [**12**].[1] By adding additional terms to the defining equation for the Sextic, the nodal singularities turn into smooth connections between pieces. This certainly ameliorates some of the problems nodal singularities cause. Particularly, nodal singularities cause infinitely small connections between pieces, which are impossible to print with any mechanical integrity. Smoothing also addresses the problems of computation coming from implicit plotters: general purpose implicit plotters, such as in Matlab, are incapable of correctly computing singularities. Because of the meshing techniques, they either produce dis- or over-connected models. Yet, the Barth Sextic is a "world record surface"; the record it attains is the theoretical maximum number of double points. Thus, smoothing away the double points seems dissatisfying. We yearn for a modeling technique that represents the singularities properly as points, yet permits printing without prohibitive difficulty.

3.1.3. *Technique: plaster.* We also find it worth mentioning plaster models, such as those found at the mathematics department at the University of Illinois at Urbana-Champaign [**13**]. The models are striking. We wanted a more modern approach, particularly since we wanted to imbue ours with electronic interactivity, requiring translucency or transparency, so plaster models are incompatible with our goals.

3.2. Our modeling technique: Numerical cellular decomposition. We required explicit computation of nodal singularities, so we turned to the *numerical cellular decomposition* algorithm. This section is not meant to be a thorough treatment of the numerical cellular decomposition algorithm for surfaces, which is discussed at length in a series of papers [**3**, **4**, **8**]. Instead, this section intends to give the reader a basic understanding of the computational algorithm which we used to compute the model we eventually turned into our physical art piece.

[1] Oliver Labs currently offers prints of these smoothed Barth Sextics for sale via Shapeways, and they are indeed stunning to behold.

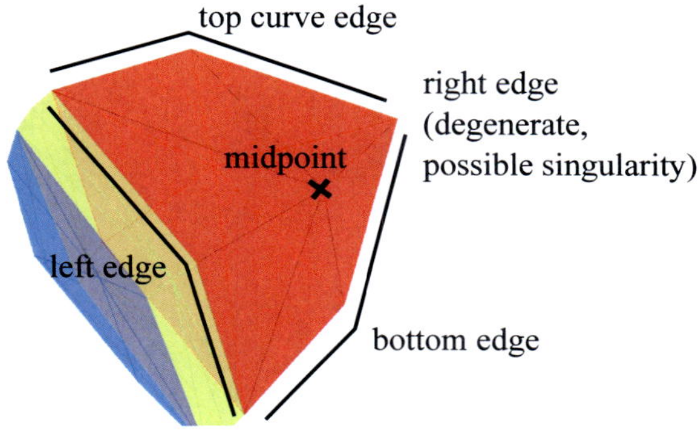

FIGURE 2. A *face* of a surface as computed by Numerical Cellular Decomposition. The red polyhedron in the center is a face of a surface representing a piece of a real surface in the numerical cellular decomposition. The face is bounded by edges of curves, which are bounded by vertices. The face has a midpoint which can be used to compute additional points.

The surface decomposition technique implemented in `Bertini_real` computes a set of *cells*, over which the implicit function theorem holds. We are working with the Barth Sextic — a two-dimensional object. In numerical cellular decomposition, a two-dimensional object is represented by a collection of *faces*, as depicted in Figure 2. The boundary of each face is an *edge* of a curve, where an edge is the cell type for one-dimensional objects. The boundary of an edge is a *vertex*, or a point, which is itself the cell type for zero-dimensional objects. This means that a cellular decomposition for a surface, regardless of the number of variables in which the object is defined, is a union of faces bounded by edges bounded by vertices. The numerical cellular decomposition is essentially an elaborate graph of connections between points in space, with a large amount of metadata.

The numerical cellular decomposition can be thought of as an implementation of the implicit function theorem. As an informal reminder, the implicit function theorem says that, except for a set of measure zero, an n-dimensional object can be represented by a union of one-to-one and onto charts between the n-dimensional unit cube, and patches on the object. Numerical cellular decomposition computes a generic smooth point in the middle of each cell, together with a homotopy which can be used to compute an arbitrary number of additional points on the cell. The decomposition algorithm also computes lower-dimensional cells representing the set of measure zero on which the implicit function theorem fails.

The main advantage of the numerical cellular decomposition for this project is that it explicitly computes the singular points on the surface, and the constructed homotopy for each face is readily translated into a triangulation of the surface. In Figure 3, pay attention to how the connections between pieces occur at the marked points. This is indicative of the structure of the decomposition; the software explicitly computes and connects the singular points. In contrast, the ray traced images in Figure 1 visually imply nodal singularities, but they are not actually computed as part of a model.

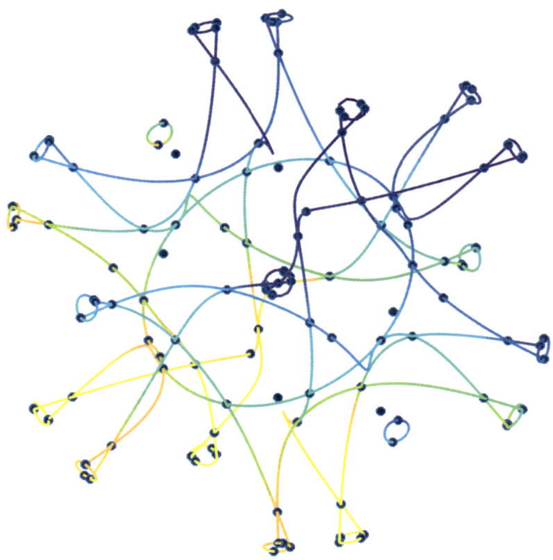

FIGURE 3. An image of the Barth Sextic generated from a numerical cellular decomposition of the surface using `Bertini_real`. The edges of the critical curve join at marked critical and singular points. In contrast to ray traced images, this model correctly resolves nodal singularities.

Note also that `Bertini_real` computes surfaces in any (reasonable) number of dimensions, so depending on the surface, we may have to project down to three coordinates, and this projection may introduce self-crossings. Fortunately, the Barth Sextic is defined in three affine variables for this project since we've set $w = 1$ in (2.1), so our triangulation from `Bertini_real` readily exports without induced crossings.

At this point we have a triangulation of the Barth Sextic and computed coordinates for each nodal singularity. In the next section, we will describe the addition of plugs and sockets at the internal singularities, as well as features for electronics.

4. Modeling and printing an assemblable Barth Sextic suitable for electronics

The software package `Bertini_real` produces `.stl` files that are suitable for 3D printing after some post-processing steps. For the model used in this paper, the Barth Sextic was decomposed into a total of 1482 faces. From these faces, one of the twenty central compact "cones" was extracted (see Figure 4) and passed through a series of post processing steps, including the use of Blender [**11**] for correcting normal vectors using their `recalculate outside` method. Notice that `Bertini_real` does not attempt to orient normal vectors for computed surfaces — they might not be orientable, and they might not even have normal vectors since the dimension of the ambient space might be greater than three. In contrast, since Blender only works for geometry in three dimensions, it handles normals.

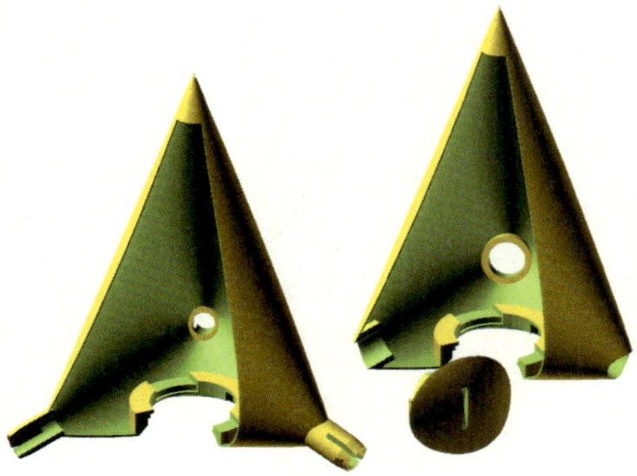

FIGURE 4. A draft of the OpenSCAD model for an assemblable Barth Sextic, with cutouts to show the internal structure. At least two sets of pieces with distinct numbers of plugs and sockets are required. The twist-in cap carries the Neopixel chip.

After producing a complete model including the singularities, we wrote code in OpenSCAD [**9**] for solidification and placement of plugs and sockets. This code is available in a public-facing git repository [**1**], separate from the code for the firmware for the light toy [**2**]. Our task was twofold:

- Provide structure that allowed the pieces to be assembled, and
- Implement design features that carry the electronics.

The problem of assemblability is solved by a plug and socket technique.[2] Additional design features were forced by our choice of components: we used Neopixel Jewel chips, a round chip carrying seven individually addressable NeoPixels. The size of the cones had to be large enough to allow the chip inside, and small enough to fit on the 3D printer platter, while accounting for print time and quality. Additionally, we wanted to hide the chips so that, at a glance, one would not suspect the piece was electronic. We settled on hollow ports between cones sufficiently large to carry the wires, together with a twist-in design for a cover carrying the LED chips.

The model design was implemented in OpenSCAD due to our prior experience and familiarity with this software, as well as the ability to parameterize features and easily put the source code under version control. We manually placed plugs and sockets at the nodal singularities of the cone extracted in Section 3.2, and angled them to permit the cones to snap together at the dihedral angle of 138.5 degrees.

Since the cones need to have internal electronics, we made the pieces hollow, with hollow connections and a service port. See Figure 4 for a screen capture from OpenSCAD of a draft of the model. In this diagram, we included a version with a cutaway to show the internal structure of the models. Note the hollow plugs, which carry wires between cones. Also shown is the coin slot twist-in cover, to which the Neopixel Jewel chips are mounted.

Note that there are two models generated for this project: one with two sockets and a plug, and another with two plugs and a socket. It is impossible to assemble a Barth Sextic using only one type of plug/socket-enabled cone, because for each

[2]We are deliberately not using the outdated cis-normative terminology *male* and *female*.

FIGURE 5. A photograph of a complete assembled snap-together electronics-capable Barth Sextic in rigid PLA plastic. We call this instance of the first design *Fixture 1*. A second instance of this design, *Fixture 2*, is nearly identical but with a different internal wiring pattern.

plug there must be a socket. For example, one cannot use a set of 20 cones each with two sockets and one plug because there would be 40 sockets but only 20 plugs. In this project, we used a model assembled from 10 cones each with one socket, and 10 cones each with two sockets. We note that there are other configurations incorporating cones with three plugs or three sockets.

In Figure 5, we present a photograph of our first completed model, which we call Fixture 1. Our second instance, Fixture 2, is a nearly identical duplicate, though the wiring between cones is a distinct pattern. Each cone is about 120mm tall, and the side walls are about 0.5mm thick, having been printed at two traces thick using a 0.25mm nozzle diameter. The total tip-to-tip diameter of the object is about 450mm, which is well larger than the printable volume of any of our group's machines. Note that one cone has a hollow torus embedded, to act as a strain relief device for hanging. Observe the wires coming from the torus.

The fragility of Fixtures 1 and 2 presents significant risk for transportation, so we adapted our original design to use a flexible plastic and only one Neopixel per cone. We turned our second design into Fixture 3. Each cone is 60mm tall, and the tip-to-tip diameter of the complete model is about 190mm. See Figure 6.

We will discuss more on implications of material choices in Section 7.1.

FIGURE 6. A photograph of a completed lit electronic Barth Sextic, printed in flexible clear TPU. We call this smaller instance of our second model *Fixture 3*.

5. Communication between controller and fixture

The interactivity of the piece is powered by a pair of Arduino Pro Mini chips communicating to each other through the I^2C protocol, built directly into the chip. The protocol requires the two communicating chips to be on the same power source, so a four-strand wire bundle connects the two pieces. We were careful in our component choices, and used Molex-style connectors, which can be repeatedly attached and removed.

The firmware for both the controller and the fixture is written in the Arduino version of the C++ programming language. A public repository for the firmware is available on GitHub [**2**]. A primary inspiration for our firmware came from a 2015 Make Magazine article on a Million Color Flashlight [**15**].

5.1. The controller's firmware. The controller was modeled in OpenSCAD, building on other models, such as an electronics base for an electronic Luxo lamp [**10, 14**]. As built it carries three potentiometers, three switches, three buttons mounted on Cherry blue keyboard switches, and a rotary encoder.

The firmware for the controller is rather simple:

```
void loop()
{
  ReadAll(); // reads the state of the inputs
  SendAll(); // sends to fixture

  delay(50); // [ms].
  // decrease to increase polling rate
  // don't want to overwhelm the fixture, though!!!
}
```

FIGURE 7. A photograph of a handwired controller featuring three potentiometers, three switches, three buttons on cherry blue keyboard switches, and a rotary encoder. It communicates with the fixture via the I^2C protocol. All three fixtures are hot-swappable.

The polling rate is set as a combination of the `delay`, and how long it takes to read and send. The large majority of the polling rate is set by the delay - however, sending input values more often likely would achieve little to no extra benefit. Note that the encoder uses an interrupt, using the `PinChangeInt` library, and its state is not sent during the loop. An improved future version of the firmware would probably use interrupts for all input devices.

The controller solely serves to send input device state to the fixture, where that data will be processed and interpreted into instructions for the LED chips in each cone. This is in contrast to a model where the controller sends visualization commands rather than raw sensor values. This would perhaps require more data to be sent over the I^2C protocol, and we appreciated the separation of concerns afforded by our implementation. By keeping data transmission small, we allow more processing power used in controlling the LED's. Given that the fixture and the controller use the same Arduino Pro Mini chip, and the lack of memory actively used in the controller, we discussed the possibility of using the I^2C connection to share memory between the two to help accommodate constraints in memory. However, we decided against this approach, due to time constraints, priorities of what we wanted to accomplish, and most importantly, the problem could likely be more readily solved simply by changing the development board used.

5.2. The fixture's firmware. The fixture's code is more complicated than the controller's code. It uses a number of global variables, as is common in the world of embedded systems. A more refined practice, perhaps using a finite state machine, would certainly clarify the code in places. However, the world of Arduino does not lend itself to the use of heavy libraries such as the C++ Standard Template Library. Nevertheless, we are happy with current results.

The fixture offers a number of modes:

- ManualRotate — The user turns the rotary encoder to drive rotational symmetries. The cones are colored by their angle from the axis of rotation, using hue in Hue Saturation Level (HSL) color space.
- AutoRotate — The fixture rotates by itself, and the user can control the speed using a potentiometer.
- XXXHighlight — The cones which form the axis of rotation are highlighted in white, and the rest are colored in hue, with saturation determined by their relative distance to the axis of rotation, so that the furthest cones from the axis are fully saturated.
- Pulse — the entire fixture pulses between 0 and 100% brightness with uniform color, where the color is determined in Red Green Blue space by the values of the three potentiometers.

The loop function for the fixture looks like this:

```
void loop() {
  getModeFromSwitches();

  switch(g_mode){ # g_mode set by controller
    case ManualRotateHighlight:{
        doManualRotate(true); break;}
    case AutoRotateHighlight:{
        doAutoRotate(true); break;}
    case ManualRotate:{
        doManualRotate(false); break;}
    case AutoRotate:{
        doAutoRotate(false); break;}
    case Pulse:{
        doPulseMode(); break;}
  }
}
```

6. Visualizing rotation using changing cone colors

The electronic interactive Barth Sextic exists both as a beautiful piece of mathematical art, and to teach about symmetry groups. In this section, we will discuss we visualize the symmetry group of the Barth Sextic using our interactive electronic model.

As the user physically rotates a digital rotary encoder, we visualize rotation of the Barth Sextic by changing the hue of each cone in hue saturation level (HSL) color space. For each rotation, certain cones cyclically trade places with each other — they are in each other's orbits. We use the orbits of each rotation to ensure that colors cycle correctly.

For each cone, we use the angle from the axis of rotation from a rotating reference angle to determine its hue. The angles of each cone were computed at compile time and written into global arrays. In contrast with cone angles, we compute the orbits at runtime — rather than storing them explicitly — due to

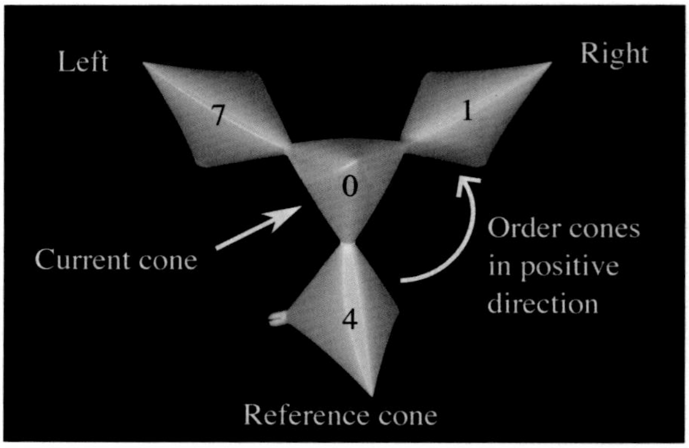

FIGURE 8. An annotated photograph of one central cone connected to three outer cones. When storing cone connections, list them in order of positive rotation. A reference cone determines which of the two remaining cones is left and right.

memory constraints. We provided modes where either the axis cones are highlighted in all white, or they rotate with the other cones.

Each of the symmetries requires knowing which cones are connected at the nodal singularities. Each cone is represented by a numeric value between 0 and 19, corresponding with the order of the electronic wiring. Cone 0 is the first in wiring order (with the hook for hanging), and cone 19 is the last one. Since there are twenty cones, and each cone is connected to three others, we hardcode a 20×3 array of integers encoding the connections between cones. We muse at the combinatorial question of how many distinct ways there are to wire the Barth Sextic, and note that Fixtures 1 and 2 used different patterns.

The order in which we encode connections matters, since we algorithmically compute cone orbits at runtime. This ordering is specified looking from the tip of a cone to the center of the fixture anti-clockwise from above. As an example, in Figure 8, cone 0 connects to cones 4, 1, and 7 in anti-clockwise order from above, and the first row of the connections is `[4, 1, 7]`.

The stored ordering is used to compute orbits in an iterative process. One common step is to determine the left or right connected cone to a central cone, using a reference cone that already appeared in an orbit, as illustrated in Figure 8.

In each orbit diagram in this section, triangles represent cones and the numbers indicate the iteration of discovery for each cone's orbit. Cones that share a number share an orbit. Additionally, the hashed gray lines indicate physical plug-socket connections between cones. In each angle diagram below, dots represent tips of cones, and angles are in degrees; sizes of dots indicate distance from the axis of rotation. When we use the word "unused", we mean a cone that has not yet been assigned an orbit number.

6.1. Two-fold rotational symmetry. The overall process for computing rotational orbits is as follows: input seed cones; determine the axis of symmetry; compute each cone's orbit and its ordering.

Two-fold rotation has 10 cone orbits, computed in seven iterations, and is visualized in the left image of Figure 9.

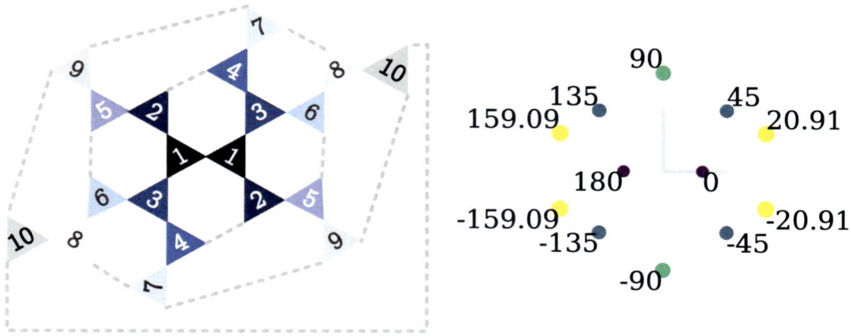

FIGURE 9. Images of the orbits and angles for two-fold rotation. In the light show, the colors of the cones are determined by their angles from a rotating reference frame as the user turns a rotatry encoder. On the left, visual representation of orbits for the two-fold rotational symmetry. There are 10 orbits, each consisting of two cones. On the right, the projection of the tips of the cones as we look down an axis of rotation, with points labeled by angle.

1. The user picks two connected seed cones using the rotary encoder and a potentiometer.
2. From each seed with the other as reference, determine the right of the non-seed connected cones.
3. From each seed with the other as reference, determine the left of the non-seed connected cones.
4. From each cone in orbit 2, determine the right unused connected cone.
5. From each cone in orbit 2, determine the left unused connected cone.
6. From each cone in orbit 3, determine the right unused connected cone.
7. From orbit 4, determine the unused connected cones.
8. From orbit 6, determine the unused connected cones.
9. From orbit 5, determine the unused connected cones.
10. From orbit 8, determine the unused connected cones.

The angles for the tips of the cones under projection along the axis of rotation determine their color in circular hue space. The angles are depicted in the right image of Figure 9. We present photographs of the piece undergoing two-fold rotation in Figure 10.

6.2. Three-fold rotational symmetry. Three-fold rotation requires eight iterations to compute cone orbits, visualized in the left image in Figure 11. Since the angles under projection along the axis of rotation determine the colors of the cones, we present the projected points with their angles in the right image of Figure 11. Here are the iterations:

1. The user picks one seed cone using the rotary encoder.
2. Find the three cones connected to the seed.

3,4. To each cone in orbit 2 there is a pair of connected cones. For each pair, find the two unused connected cones which form a path of length four. The left cones form orbit 3, and the right cones form orbit 4.

FIGURE 10. A pair photographs of electronic visualization of two-fold rotation. The two central white cones indicate the axis of rotation, through the nodes joining two antipodal pairs of cones. In the left photo, a red cone is on top. In the right, the red cone has moved to the bottom, 180 degrees of rotation.

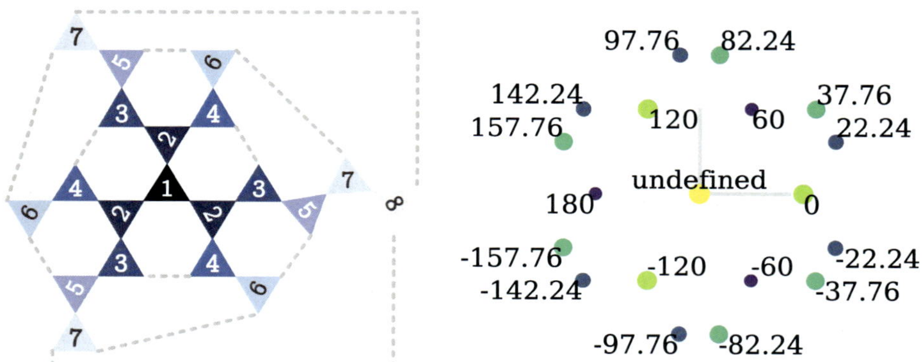

FIGURE 11. Orbits and angles for three-fold rotation. On the left, visual representation of orbits for the three-fold rotational symmetry. There are eight orbits, two with one cone each, and six with three cones each. On the right, the projection of the tips of the cones as we look down an axis of rotation, with points labeled by angle in degrees.

5,6. Repeat steps 3, 4 with pairs from orbits 3 and 4. Found cones connected to orbit 3 form orbit 5, and those connected to orbit 4 form orbit 6.
7. Determine the mutually connected unused cones between each pair from orbits 5 and 6.
8. The last orbit is the one remaining unused cone.

We present photographs of the piece rotated 120 degrees in Figure 12.

6.3. Five-fold rotational symmetry. The orbits for five-fold rotation requires only four comparatively simple iterations, visualized in Figure 13.

FIGURE 12. Three photographs of electronic visualization of three-fold rotation. One central white cone indicates the axis of rotation, through the tips of two antipodal cones. The red cone points outward in each photograph, as the colors of the cones rotate 120 degrees through a complete revolution.

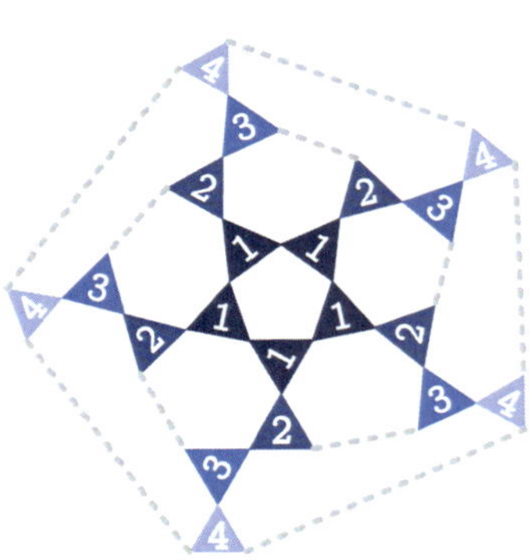

FIGURE 13. Visual representation of orbits for the five-fold rotational symmetry. There are four total orbits, with five cones in each.

1. The user picks two seed cones using the rotary encoder and a potentiometer; always going right picks three more cones, forming a pentagonal loop of length five.
2. Find the unused cones connected to each cone in orbit 1.
3. Find mutually connected cones between each adjacent pair in orbit 2.
4. Find the unused cones connected to each cone in orbit 3.

For determining color of the cones in hue space, we project tips of the cones and they form a pair of aligned regular pentagons. Since this is geometrically simple compared to two- and three-fold rotation, we did not provide an image of this

FIGURE 14. Five pentagonally arranged photographs of electronic visualization of five-fold rotation. Five cental white cones indicate the axis of rotation, through the center of a fictitious pentagonal face. A red cone points directly out from the center in each rotation step of 72 degrees.

projection or their angles. In Figure 14 we present five photographs of the fixture undergoing five-fold rotation.

7. Considerations when going electric

While this paper is largely about the technical aspects of modeling, printing, and programming our interactive Barth Sextic, our underlying motivation is to invite you to incorporate electronics into your projects. Make your projects move, make them modular, make them dynamic. This section discusses some of the considerations, choices, and realizations we made during completion of our electronic Barth Sextic project. We hope they are helpful to others engaged in similar activity.

7.1. Material choice. Material choice becomes very important especially if you allow people to handle your printed pieces. Accidents are prone to happen in one way or another. If shipping pieces, there is a chance the piece will break during transit. If showing someone else, they could drop the model. Even as researchers we are human and could knock a piece over. Material choice helps control loss — choose a material compatible with your vision for how you will share and display your piece.

Note that this section is not an exhaustive comparison of material choices. Rather, we focus on those for fused filament fabrication (FFF), since that is what we practice. Our group insists on printing our own pieces, deeming other printing techniques either too expensive (powder printing, etc.), or too risky to personal health to be in an office setting (resin printing with SLA).

FIGURE 15. A photograph of six Barth Sextics arranged from large to small. The largest is about 190mm in diameter; the smallest is an earring. From left to right, they are printed from glow-in-the-dark PLA, purple TPU, glow-in-the-dark PLA, purple PLA, white TPU, and white TPU. All were printed in-office using FFF technology.

Polylactic acid (PLA) is a plastic extremely common in consumer 3D printing. It is inexpensive, does not smell too bad when printing, and prints reliably. However, PLA has poor heat properties. With a softening temperature as low as 50-60°, it will melt or warp in a car on a sunny day, during shipping in the heat of summer, and when used in proximity to electronics. For example, one of the instances of the Sextic in this paper was kept in a box of styrofoam peanuts during code development by a student who did not know better. The heat built up in the box and most of the cones warped. This cannot be repaired without reprinting, together with manual dis- and re-assembly of the electronics. Furthermore, we warn that PLA is brittle and breaks easily. PLA is a poor choice for fragile pieces requiring shipping or show without the protection of a display case.

We would also like to warn that PLA is not very workable, compared to, say acrylonitrile butadiene styrene (ABS). If one tries to use a drill to make holes, or sandpaper to smooth or remove material, PLA is prone to melting. This means that large burrs develop and the material scars, often tearing along trace lines. ABS on the other hand drills quite nicely, sands well, and is amenable to various glues and paints such as those for model cars and miniatures.

Fortunately, there are a variety of alternatives to PLA, with various advantages and disadvantages. The plastic we have gravitated towards is thermoplastic polyurethane (TPU). This plastic is extremely durable, capable of sustaining connections as small as a few toolpaths, down to 0.5mm total in cross-sectional area. Any other FFF-compatible plastic would simply break at this scale. TPU is also durable to dropping and impact, so it is well-suited for fragile pieces which must be shipped. In addition, its durability lends itself to small-scale printing, e.g. for earrings and other jewelry. See Figure 15 for photographs of six Barth Sextics, including three office-printed TPU pieces. In particular, note the incredibly small connections between cones in the purple Sextic, second from the left — this piece took about 130 hours of print time (and spurred the purchase of an uninterruptible power supply). If those were printed in a rigid plastic, they would likely break upon removal of support material.

We would like to add that TPU is a great material for models which require support. Support printed in the same material as the model removes fairly cleanly, and even large amounts of support can be removed easily using only pliers, clippers, and a hobby knife (strop regularly!). TPU also can be supported by polyvinyl alcohol (PVA) pushed through a second extruder. While dual-extrusion does produce cleaner prints with less time wasted on manual support removal, it comes with its own challenges. One primary challenge of dual extrusion is the doubling of exposure to various mechanical failures, simply because it doubles the number of extruders.

A challenge specific to PVA is that it crystallizes when held at extrusion temperature, meaning that long layer times become risky; in turn, this means that small objects are tractible, while large prints are prohibitively scary. Furthermore, PVA is ruined by moisture from the air. Keeping it dessicated, as well as drying it after moisture exposure, are troublesome. There are other support material choices, but there's no free lunch. Our conclusion: at-home or in-office FFF dual extrusion is currently not worth it.

A few notes on our chosen materials and machines. We printed the cones for Fixtures 1 and 2 from eSun Warm White 1.75mm PLA plastic, and the assembled model is incredibly fragile. Each cone took about six hours to print, meaning that it takes about a week to print a complete object. We used a heavily modified Rostock Max V2 delta printer from SeeMeCNC, with a clean PEI sheet for bed adhesion. We learned valuable lessons from Fixtures 1 and 2. In particular, the PLA models, beyond being brittle, are susceptible to heat warping even from the little heat of the LEDs.

We used flexible TPU for fixture 3, and it is far more durable. We used a 0.25mm nozzzles to print SainSmart Clear TPU 1.75mm plastic onto a PEI bed coated with glue stick to prevent extreme adhesion problems. For this iteration, we used a stock Ultibots D300 VS delta printer.

7.2. Electronic components. The choice of development boards for your project has enormous impact on capabilities, development, and deployment.

In our electronic Sextics, we chose to use the Arduino Pro Mini. We used the 5V model so that we did not have to worry about using changing voltage between the Arduino, and other components including the LEDs.

The Arduino Pro Mini carries a small amount of memory (in fact, downright miniscule by standards of modern personal computers), both for program instructions and variables.[3] Working around memory limitations restricted our programming approach to memory-lean paradigms. Our initial approach was to use a `stack` of light states, but it made the chip go out-of-memory, leading to undefined behaviour and weird hard-to-predict bugs. We had to scrap the `stack`, and move to a before-and-after two-state model, where we store the LED color for each cone in a `to` and `from` state, and interpolate between them with a fixed time interval. This severly limits the kinds of light shows that can be performed. More elaborate shows, where each cone can be interpolated or controlled independently, readily exceed the memory limits of the Pro Mini, even with significant compression techniques.

[3]We used the Arduino Pro Mini, on the ATmega328 microcontroller. This board has 2KB of SRAM, 1KB of EEPROM, and 32KB of flash memory. Compare to the Mac on which the first author is writing this paper: 40GB of RAM, and practically infinite disk space.

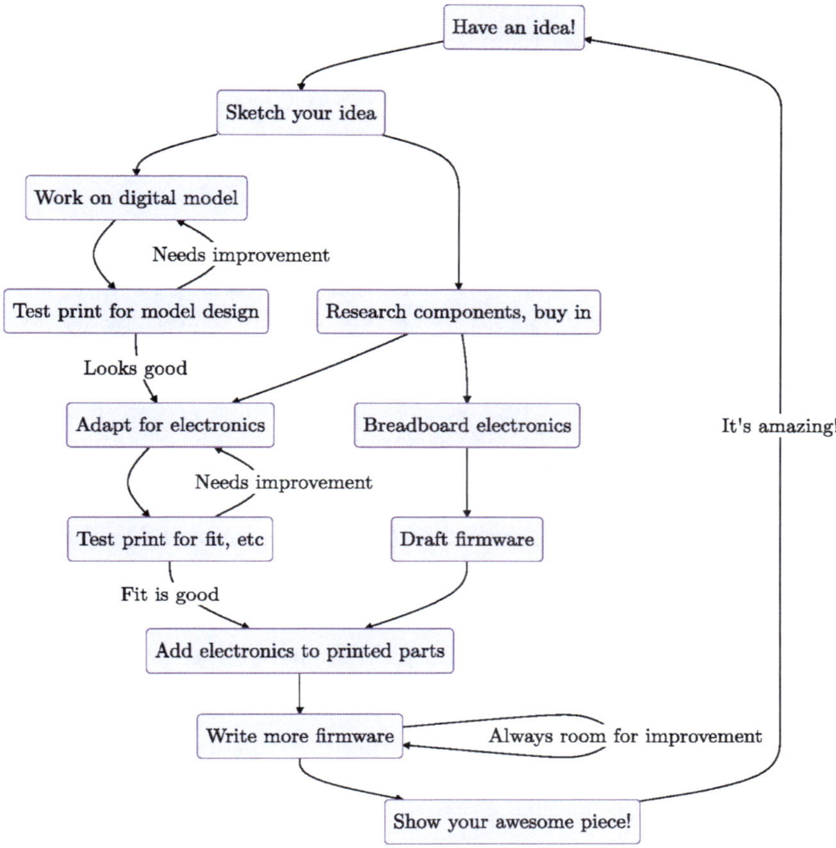

FIGURE 16. A flowchart describing a general process for adding electronics to 3D printed projects. We note two parallel but intertwined tracks for developing a physical model and firmware.

Another board perhaps would have been a better choice, allowing for a tighter loop, smoother animation, and more complicated visualizations. Due to the authors' prior experience with the Pro Mini, its on-board capabilities for I^2C communication when paired with another Pro Mini, and cheap pricetag (about \$7 each at the time of purchase in 2019), the Pro Mini was an easy choice.

We envision the next version will use something like an ESP8266. This would allow significantly more complicated shows, wireless remote controlling with a virtual interface, and perhaps even environmental interaction.

7.3. A mindset for adding interactivity. We would like to suggest a process and practice for incorporating electronics into a 3D printed object. We present a flowchart attempting to illustrate the steps in our process, which we hope you find helpful in thinking about your own projects. Please see Figure 16.

A new project starts with an idea. Sketching on paper is tremendously valuable, as the digital design process can be very difficult, often involving the invention of new techniques and creation of new mathematics!

Since the idea involves electronics, there are two tracks to pay attention to: physical modeling and firmware implementation. The model must serve the mathematical concept to be illustrated, while being printable and durable enough to be handled for assembly. It must also be able to carry the components, all without sacrificing aesthetic appeal. Simultaneously, the electronics design must correspond to the interactive part of the idea being illustrated, and can be hard to prototype until a significant part of the design process is completed.

Certainly, the two development tracks influence each other. The physical model must be designed with the specific components in mind. In our project, we had to ensure that the plugs between cones had sufficient clearance for three strands of wire to pass with minimal pulling. If we were to design a third version, we would incorporate a few wire guides into the bottom of the cone to keep wires from in front of LEDs. We also had to model the twist-in service port to carry the specific NeoPixel Jewel chip in the first PLA version. In the flexible version we adapted the port to carry a single-pixel chip. The electronics profoundly influence the design, and must be incorporated early in the process to avoid repeated design changes and costly prototyping (both in materials and time).

We encourage the reader to expect to print many drafts and versions. Proper mechanical fit can be difficult to achieve on the first attempt. Material properties can also cause revelations in thought during the design process.

A prototyping step is important to getting the firmware authorship process off to a good start. It lets the maker ensure that they have the correct components on hand before final assembly, which can be difficult to revise. In our piece, assembly was done with a soldering iron by hand. To replace components would require de-soldering in close proximity to plastic — not a good combination. Prototyping also is good practice for getting a first version of the firmware up and running, to ensure basic functionality at time of assembly.

We have one more note about firmware: use version control software. Maybe it is obvious to the reader, but in the context of undergraduate research projects, version control is incredibly important. Not only is version control a great marketable skill for our students, but also it solves bothersome problems related to collaborative software authorship, such as nonsensical emailing of "hot copies" of code files. Furthermore, use of `git` in particular allows for commits in software to track who made the contributions, leading to a public system for crediting work on a project (assuming the repository is publicly available). Finally, we think that version control with public repositories is a great way to share work with the general public and empower other communities to use our work and make our projects. Both the model and the firmware for the project described in this paper have public repositories on Github [**1**, **2**].

8. Mathematical and technical skills fostered by this project

The first author of this paper teaches at an undergraduate liberal arts university. The second and third were undergraduate students for this research. High impact practices and undergraduate collaborative research are critical elements of our instructional programs. This multi-year interactive electronic mathematical art project involved:

- Group theory, in understanding the symmetry group of the Barth Sextic.

- Mathematical software authorship, in the algorithm to compute the surface, in generating the printable digital model, in using Python to producing the STL model from the mathematical software `Bertini_real`, and in encoding the graph structure of surface for the LEDs.
- Computer programming skills, in using C++ to make the fixture communicate with the controller, in making the LEDs display the correct colors, and designing interesting and instructive modes of operation.
- Advanced computer science skills, in working around the significant space constraints of the Aduino's low-memory environment.
- Version control software, both for controlling progress in the software, and solving the hot-copy problem that inevitably arises in these kinds of projects.
- Crafting skills, in operating the 3D printer, including slicing and post-processing the objects.
- Photography skills, in capturing still photos and videos of the object during development and after completion.
- Presentation skills, in showing the piece to the public at art shows and exhibits.

Adding electronics to your mathematical art project does much more than add interactivity. Involving undergraduate students in your hybrid project will support interdepartmental research beyond the mathematics department, and foster skills sought after by many employers. If your university or school has a maker space or 3D printer lab, this will help your students use more than just the printers, but also the soldering station, the coding environment, and interfacing with the public.

By designing beyond the build platform, we gain access to a world of projects for show, teaching, and mentoring.

Acknowledgments

The first author wishes to thank all the participants of the Illustrating Mathematics semester program at the Institute for Computational and Experimental Research in Mathematics in Fall 2019. Many formative conversations spurred the ideas implemented in this paper.

We thank aꙒa for his help in complying with the AMS Style Guide, though we do wish modern academic writing style permitted the use of contractions.

The constructive criticism provided by the referee was helpful in focusing, organizing, and clarifying our ideas. Thank you.

References

[1] silviana amethyst, *Sextic lamp*, `github.com/amethystlab/sextic_lamp`, 2022.

[2] silviana amethyst, *Sextic light toy*, `github.com/amethystlab/sexticlighttoy`, 2022.

[3] silviana amethyst, Daniel J. Bates, Wenrui Hao, Jonathan D. Hauenstein, Andrew J. Sommese, and Charles W. Wampler, `Bertini_real`*: software for one- and two-dimensional real algebraic sets*, Mathematical software—ICMS 2014, Lecture Notes in Comput. Sci., vol. 8592, Springer, Heidelberg, 2014, pp. 175–182, DOI 10.1007/978-3-662-44199-2_29. MR3334763

[4] silviana amethyst, Daniel J. Bates, Wenrui Hao, Jonathan D. Hauenstein, Andrew J. Sommese, and Charles W. Wampler, *Algorithm 976:* `Bertini_real`*: numerical decomposition of real algebraic curves and surfaces*, ACM Trans. Math. Software **44** (2017), no. 1, Art. 10, 30, DOI 10.1145/3056528. MR3685393

[5] The authors of Surfer, *Surfer version 1.7.0 (100)*, `www.imaginary.org/program/surfer`, 2010.

[6] John Baez, *Barth Sextic*, `blogs.ams.org/visualinsight/2016/04/15/barth-sextic/`, 2016.
[7] W. Barth, *Two projective surfaces with many nodes, admitting the symmetries of the icosahedron*, J. Algebraic Geom. **5** (1996), no. 1, 173–186. MR1358040
[8] Daniel J. Bates, silviana C. amethyst, Jonathan D. Hauenstein, Andrew J. Sommese, and Charles W. Wampler, *On computing a cell decomposition of a real surface containing infinitely many singularities*, Mathematical software—ICMS 2014, Lecture Notes in Comput. Sci., vol. 8592, Springer, Heidelberg, 2014, pp. 246–252, DOI 10.1007/978-3-662-44199-2_39. MR3334773
[9] Maintained by Marius Kintel, *Openscad version 2021.01*, `openscad.org`, 2021.
[10] DanielBull, *Large Luxo Jr*, `www.thingiverse.com/thing:432648`, 2014.
[11] The Blender Foundation, *Blender version 3.0.1*, `blender.org`, 2022.
[12] Oliver Labs, *Barth's sextic*, math-sculpture.com/barth-sextic/, 2022.
[13] Department of Mathematics, *The altgeld math models*, `www.mathmodels.illinois.edu/cgi-bin/cview?SITEID=4&ID=342`, 2022.
[14] ofloveandhate, *Electronics Base for Large Luxo Jr.*, `www.thingiverse.com/thing:1996656`, 2016.
[15] Dan Rasmussen, *Million color hsl flashlight*, Make Magazine **46** (2015), 64–67.
[16] E.W. Weisstein, *Barth sextic*, `mathworld.wolfram.com/BarthSextic.html`, 2022.

Department of Mathematics, University of Wisconsin - Eau Claire, Eau Claire, Wisconsin 54702
Email address: `amethyst@uwec.edu`

Department of Mathematics, University of Wisconsin - Eau Claire, Eau Claire, Wisconsin 54702
Email address: `SamanthaLMaurer@gmail.com`

Department of Mathematics, University of Wisconsin - Eau Claire, Eau Claire, Wisconsin 54702
Email address: `william.tom.obrien@gmail.com`

Proceedings of Symposia in Applied Mathematics
Volume **79**, 2023
https://doi.org/10.1090/psapm/079/00712

Deformation spaces and static animations

Gabriel Dorfsman-Hopkins

ABSTRACT. We study applications of 3D printing to the broad goal of understanding how mathematical objects vary continuously in families. To do so, we model the varying parameter as the vertical axis of a 3D print, introducing the notion of a *static animation*: a 3D printed object each of whose layers is a member of the continuously deforming family. We survey examples and draw connections to algebraic geometry, complex dynamics, chaos theory, and more. We also include a detailed tutorial (with accompanying code and files) so that the reader can create static animations of their own.

1. Introduction

Across mathematical disciplines there is considerable interest in understanding how a mathematical object can deform according to a shift in a single parameter, producing a varying family of related mathematical objects. If these objects have small enough dimension, one can vary the parameter over time and display how the object changes as an animation on a screen. With a 3D printer we can replace the time dimension with the vertical z-axis, and stack each successively deformed layer on top of the last. This turns the entire parametrized family of objects into a single solid which we can hold in our hands, giving an entire new physical dimension to the deformation.

The study of continuously varying families is ubiquitous in mathematics. To name a few examples: In algebraic geometry, one studies degenerations of curves, surfaces, and higher dimensional algebraic varieties, putting them in continuously varying *flat families* which interpolate between one space and the next [**13**, Chapter III.9]. In complex dynamics, one asks about limiting behavior when iterating a holomorphic function f, and compares this to the limiting behavior of a slight perturbation of f, to dazzling effects [**17**]. In topology, the question of whether one space can continuously deform to another forms the notion of a homotopy, defining the fundamental invariant in the rich and expansive field of homotopy theory [**14**, Chapter 4]. In the study of chaotic dynamical systems, one explores how slight changes in initial conditions can radically affect longer term outcomes [**24**]. In fact, one would be hard pressed to find a field of mathematics which does not contain a multitude of problems that aim to understand how a mathematical object or system varies when perturbed.

2020 *Mathematics Subject Classification.* Primary 00-01, 65D17, 68U05, 37F75.

In this article we propose a general framework which can be used to explore situations like those described in the previous paragraph. Namely, we describe ways to unify an entire family of continuously varying objects into a single mathematical object or *deformation space*. When the members of this family are 1 or 2 dimensional, we argue that 3D printing this deformation space provides new perspectives on the illustration of continuously deforming geometry, beyond that which an animation on a screen can achieve. We call these 3D models *static animations*.

There are two main components to this article, one more theoretical and one more interactive. The theoretical component precisely formulates the notion of a static animation and describes connections to various mathematical areas, showcasing examples and suggesting avenues for exploration. The interactive component provides detailed instructions for the reader to create static animations of their own, and is accompanied by code and files in a public github repository [**5**].

The structure of the article is therefore as follows. In Section 2 we give a rigorous definition of the notion of a static animation, and illustrate the concept with an example of a varying family of polar flowers. In Section 3 we introduce the author's first 3D printed deformation space, which illustrates a family of Pythagorean tree fractals. In Section 4 we enter the world of complex dynamics, showcasing various prints of deformations of Julia sets, including some our own work, as well as work Bernat Espigulé, Caroline Davis, and Bethany Mussman.

We then arrive at the interactive portion in Section 5, describing a procedure to create static animations that was developed with Bernat Espigulé. This workflow can be understood as a sort of *reverse tomography*. Tomography is a way to visualize the internal structure of 3D object by producing a collection 2D pictures of cross sections at various depths. For example, an MRI is a type of tomography which can be applied to the human body. The process we describe goes in the other direction, starting with a collection of 2D images and stitching them together into a 3D object.

The upshot of this particular workflow is that the user can create static animations using mostly 2D modelling methods—minimizing the need for more difficult 3D modeling software. The main idea is to feed a stack of 2D images to an open source medical and molecular visualization software called Chimera [**21**],[**25**] which will stitch them together into a 3D model.[1] After broadly describing the workflow, we also include a detailed step-by-step tutorial allowing the reader to follow along and create a static animation of their own. Accompanying the tutorial we have our own code and files from each step of the process, including a file of the finalized 3D model. It is the hope of the author that after working through this tutorial, any reader able to make 2D animations on a screen will also be able to make 3D static animations.

The idea of viewing a continuously deforming family of a geometric objects as a geometric object in its own right is certainly not new, and in fact it was the algebrogeometric perspective of viewing so-called flat morphisms as parameter spaces which motivated the author's exploration into this field. We provide more details and an example of this connection in the appendix.

[1] As far as the author is aware, this functionality was initially developed to reconstruct 3D models from medical tomographical images such as MRI images.

2. Static animations

In what follows we give a precise mathematical formulation of what we mean when referring to a static animation.

DEFINITION 2.1 (Static animations). Suppose we are given some interval $I \subseteq \mathbb{R}$, and for each $t \in I$ we are given a subset $B_t \subseteq \mathbb{R}^2$. The *static animation* associated to the data $(I, (B_t)_{t\in I})$ is the following subset of $\mathbb{R}^3$.

$$\mathcal{B} = \{(b,t) : t \in I, b \in B_t\} \subseteq \mathbb{R}^3.$$

The sets B_t are called the *frames* of the static animation.

This gives us some subset of $\mathbb{R}^3$, which we could hope to 3D print.[2] Observe that a static animation comes equipped with a projection map $\pi : \mathcal{B} \to I$ given by the rule $\pi(b,t) = t$. For any $t_0 \in I$ we can recover the frame B_{t_0} as the the fiber of π over t_0:

$$\pi^{-1}(t_0) = \{(b,t_0) \in \mathcal{B}\} \cong B_{t_0}.$$

We illustrate the notion of a static animation and its associated projection map with an example of a continuously varying family of curves, interpolating between the polar flowers B_{t_1}, B_{t_2}, and B_{t_3} displayed in Figure 1. We then align these

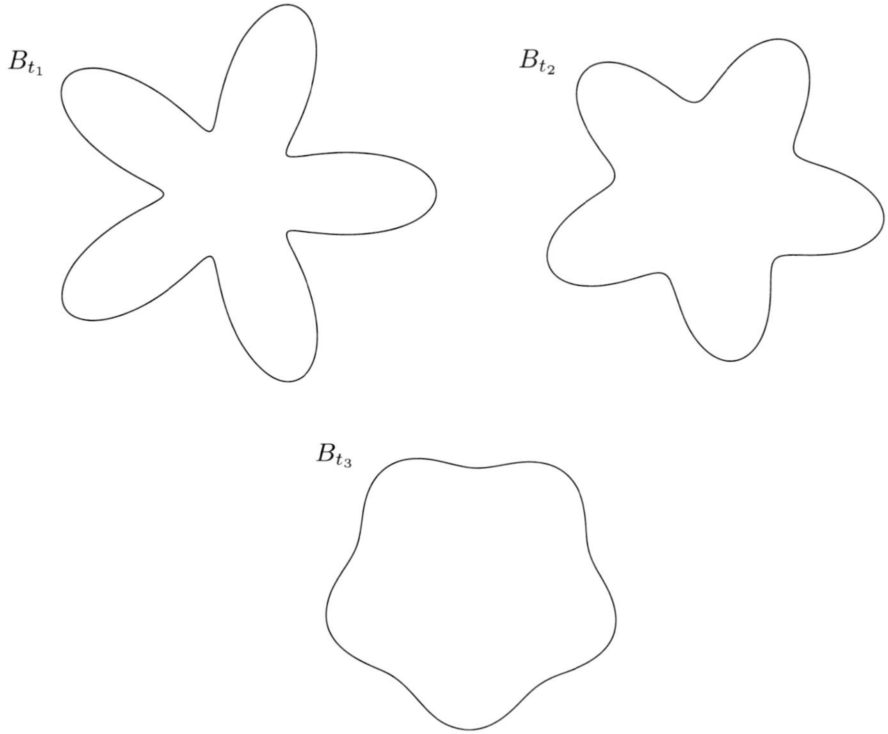

FIGURE 1. Polar flowers which will serve as frames for a static animation.

[2]That said, if the set B_t varies too wildly with t, we will probably be unable to do so. To have any hope, we'd like B_t to vary continuously in t. Loosely speaking, this means that if $t \approx t'$ then $B_t \approx B_{t'}$. This can be made precise in a number of ways, depending on the context.

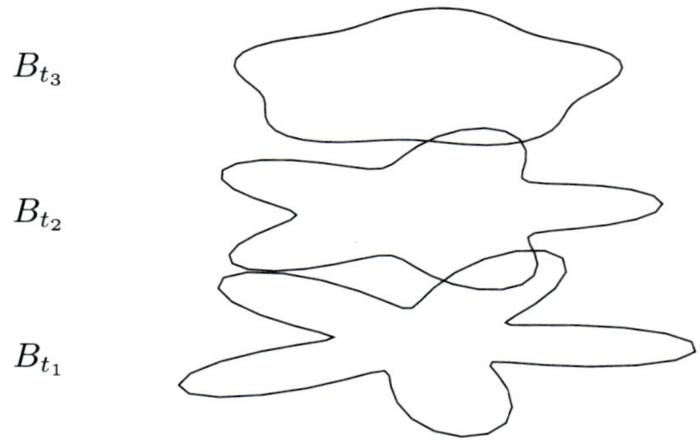

FIGURE 2. Frames of a static animation, aligned vertically.

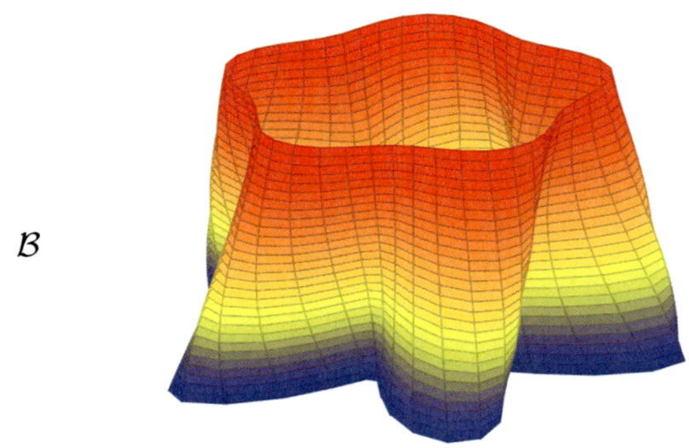

FIGURE 3. Interpolating between the discrete frames of a static animation to make a surface in $\mathbb{R}^3$.

curves vertically, with the polar curve B_t curve living on the plane $z = t$ (Figure 2). Finally, we interpolate by filling in the rest of the t values to get the surface, $\mathcal{B}$, (Figure 3). We will explicitly describe this surface in Section 2.1 below.

The projection $\pi : \mathcal{B} \to \mathbb{R}$ consists of simply plucking out the z-coordinate of a point on this surface. For $t \in I$, the fiber over t of this projection, $\pi^{-1}(t)$, is therefore the horizontal cross section given by intersecting $\mathcal{B}$ with the plane $z = t$. By construction, this is precisely the polar flower B_t, floating at height t.

We can now cap this surface to make a solid deformation space illustrating how to continuously interpolate between polar flowers (Figure 4), and send this model to a 3D printer and obtain the physical object (Figure 5). We give a step by step tutorial for the creation of this object in Section 5.2. An STL file of the 3D model is also available at our public github repository [**5**], and can be downloaded to print.

Of course, one could also illustrate the continuous deformation of polar flowers as an animation on a screen.[3] Indeed, the horizontal cross sections of the solid described above correspond precisely to the frames of this animation. This is why we call these objects *static animations*, and their horizontal cross sections *frames*.

2.1. A mathematical description of the deformation of polar flowers. We continue with a more precise formulation of the example above. Adopting the

[3]We made such an animation. Find it here: `http://www.gabrieldorfsmanhopkins.com/digital_fabrication/polarFlowerDeformation/flowerAnimation.gif`

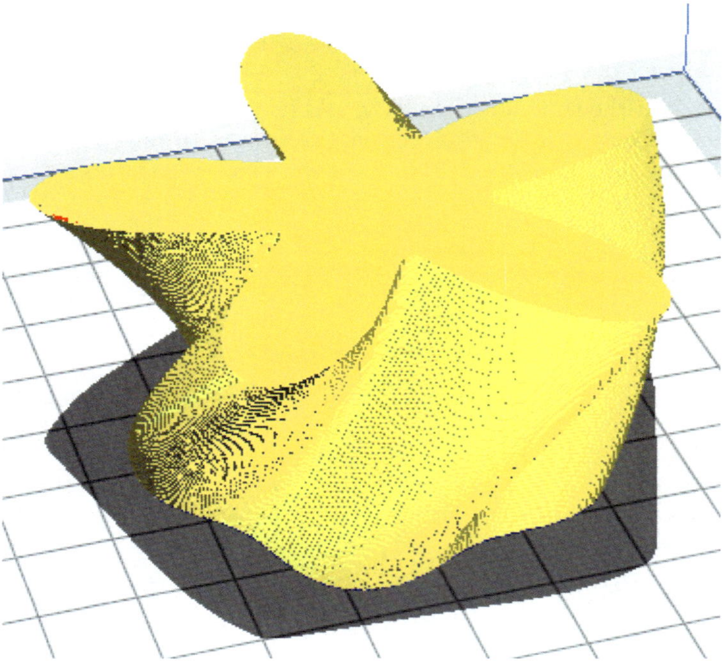

FIGURE 4. The static animation, converted into a 3D printable solid.[4]

FIGURE 5. The 3D printed static animation of deforming polar flowers.

same notation, B_{t_1} is the polar flower given by the equation

$$r = 2 + cos(5\theta)$$

in polar coordinates. Similarly, B_{t_3} is given by

$$r = 2 + \frac{1}{5}\cos(5\theta + 2\pi/5).$$

[4]This is a screenshot of the STL file opened and rendered in Ultimaker-Cura [**4**]

In particular, both consist of waves on the circle of radius 2, although they differ in amplitude and a shift of where the wave begins. These correspond to two varying parameters in the equation: the amplitude is controlled by the coefficient in front of the cosine, and the shift corresponds to adding a constant term to the argument of the cosine. These can be adjusted simultaneously to interpolate continuously between B_{t_1} and B_{t_3}, giving a family of curves whose general element B_t is given by the equation

$$r = 2 + t\cos(5\theta + 2\pi t).$$

Indeed, letting t=1 we recover B_{t_1} and letting $t = 1/5$ we recover B_{t3}.[5] From this we derive an explicit parametrization for our deformation surface

$$\rho : [0, 2\pi] \times [.2, 1] \longrightarrow \mathbb{R}^3.$$

Giving target cylindrical coordinates, ρ can be expressed by the rule

$$\rho(\theta, t) = (2 + t\cos(5\theta + 2\pi t), \theta, t).$$

The static animation $\mathcal{B}$ is precisely the image of ρ, and by restricting to $\mathcal{B}$ the projection $\mathbb{R}^3 \to \mathbb{R}$ onto the z-axis, we recover the projection $\pi : \mathcal{B} \to I$. For $t_0 \in I$ we compute that

$$\pi^{-1}(t_0) = \{\rho(\theta, t_0) : \theta \in [0, 2\pi]\} = \{(2 + t_0\cos(5\theta + 2\pi t_0), \theta, t_0) : \theta \in [0, 2\pi]\}.$$

This is precisely the polar flower given by $r = 2 + t_0\cos(t\theta + 2\pi t_0)$—that is, the frame B_{t_0}—floating at the fixed height t_0.

2.2. Higher dimensional stacks of frames. It is worth returning to the observation that there were two varying parameters in our family of polar flowers, one changing amplitude and another controlling the phase shift. We controlled both using the single parameter t, but of course each can move independently, say, with parameters s and t. This would give a 2-dimensional parameter space of polar flowers, say $\mathbb{R}^2$, and for every $(s, t) \in \mathbb{R}^2$ we could consider the curve

$$C_{s,t} : \quad r = 2 + s\cos(5\theta + 2\pi t). \tag{2.1}$$

This puts us in a situation similar to Definition 2.1, except that we have a 2-dimensional stack of frames. Therefore, trying to build the static animation would put us in 4-space,

$$\mathcal{C} = \{(b, s, t) : b \in B_{s,t}\} \subseteq \mathbb{R}^2 \times \mathbb{R}^2 = \mathbb{R}^4.$$

$\mathcal{C}$ can also be explicitly parametrized, via the rule:

$$(\theta, s, t) \mapsto (r, \theta, z, w) = (2 + s\cos(5\theta + 2\pi t), \theta, s, t) \subseteq \mathbb{R}^2 \times \mathbb{R}^2,$$

where the first $\mathbb{R}^2$ is given polar coordinates. Of course, this attempted static animation is now a 3-dimensional hypersurface in $\mathbb{R}^4$, making it impossible to 3D print. This is exactly the type of situation studied in Section 4 where we look at deformations of Julia Sets which have a 2-dimensional parameter space (the Mandelbrot set), so rather than just give up here, we will briefly describe a way to cut down a dimension.

Let I be an interval and consider a path $\gamma : I \to \mathbb{R}^2$, which we express coordinatewise as $\gamma(t) = (x(t), y(t))$. Then for every $t \in I$ we can consider the frame

[5]We can also recover B_{t_2} by letting $t = 3/5$.

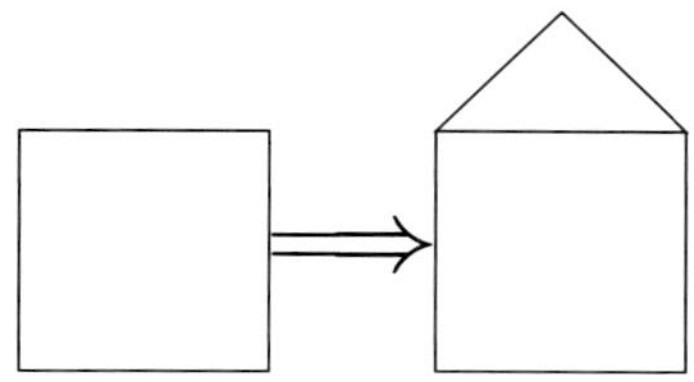

FIGURE 6. The substitution rule for the Pythagorean Tree Fractal.

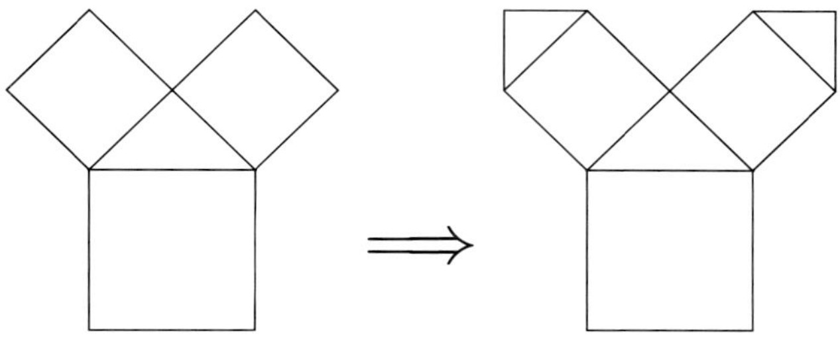

FIGURE 7. Iterating the substitution rule for the Pythagorean Tree Fractal.

$D_t := C_{x(t),y(t)}$ defined as in Equation (2.1). We are now back in the setup of Definition 2.1 and can form the static animation

$$\mathcal{D}_\gamma := \{(d,t) : t \in I, d \in D_t = C_{x(t),y(t)}\} \subseteq \mathbb{R}^3.$$

For example, if we let $\gamma : [.2, 1] \to \mathbb{R}^2$ be the path $\gamma(t) = (1-t, 1-t)$ (consisting of travelling from $(1, 1)$ to $(.2, .2)$ in a straight line along the diagonal), then $\mathcal{D}_\gamma$ recovers the static animation $\mathcal{B}$ from Section 2.1.

The key takeaway is the following: if one has a continuously varying family of 2D objects which vary over a higher dimensional parameter space P, one can still create 3D static animations by considering one parameter variations associated to paths within P.

3. Deforming the Pythagorean tree fractal

The Pythagorean tree fractal is a recursive plane fractal, constructed as follows. First, one attaches the the hypotenuse of a right triangle to the side of a square (Figure 6). Two squares are then attached to the legs of this triangle, and the process repeats, with the hypotenuse of a right triangle affixed to the opposite side of each new square (Figure 7). Squares are added to the legs of these triangles, and the process continues recursively forever (Figure 8). Notice all of the triangles in the construction above were congruent, in particular they were isosceles right triangles with acute angles of 45°. One could ask what happens if one used, say, a 30-60-90 triangle at each step instead. As before, one begins by aligning the hypotenuse of a 30-60-90 triangle with the top of the square (Figure 9). Then, following a similar process, two squares are attached to the legs of this triangle, although in this case they will not be the same size, instead matching the two differently sized legs. Continuing, the hypotenuse of an appropriately sized 30-60-90 triangle is attached to the opposite side of each square, making sure to maintain consistent orientations (Figure 10) As before, we continue this process recursively, this time observing that

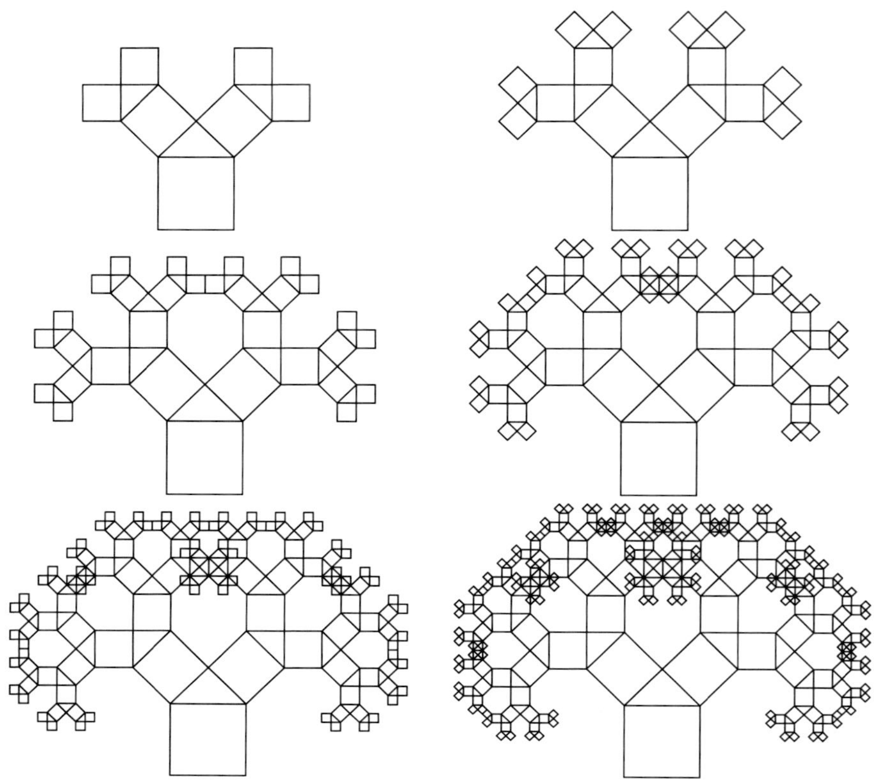

FIGURE 8. More iterations of the substitution rule, giving rise to an apparent global structure.

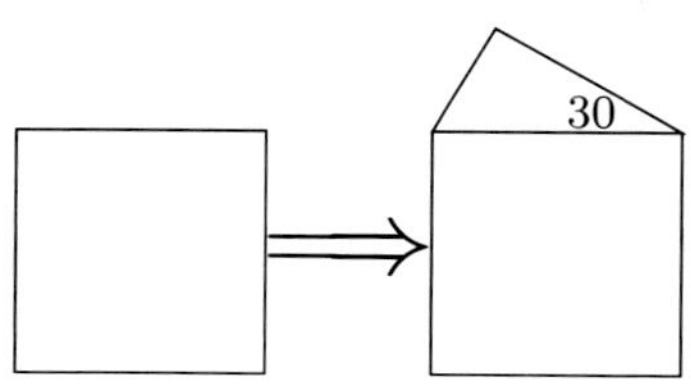

FIGURE 9. The substitution rule for a Pythagorean Tree Fractal based on a 30-60-90 triangle.

the global effect is somewhat different than when we used isosceles triangles (Figure 11).

A right triangle is determined (up to congruence), by its smallest angle θ. In particular, for each angle $0 < \theta \leq 45$, one can follow this construction using a right triangle whose acute angles are θ and $90 - \theta$. By varying the angle θ, one obtains a varying family of planar fractals parametrized by θ. In Figure 12 we display the resulting fractal for some values of θ between 45 and 30. The images seem to indicate that these fractals interpolate between the two that we have seen so far, suggesting the creation of a (static) animation.

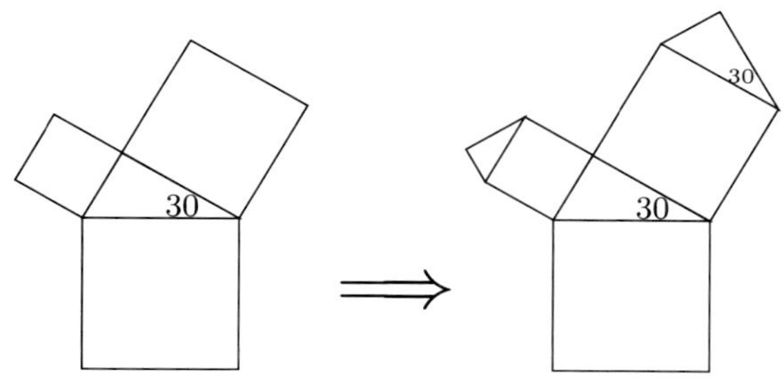

FIGURE 10. Iterating the substitution rule for the 30-60-90 Pythagorean Tree Fractal

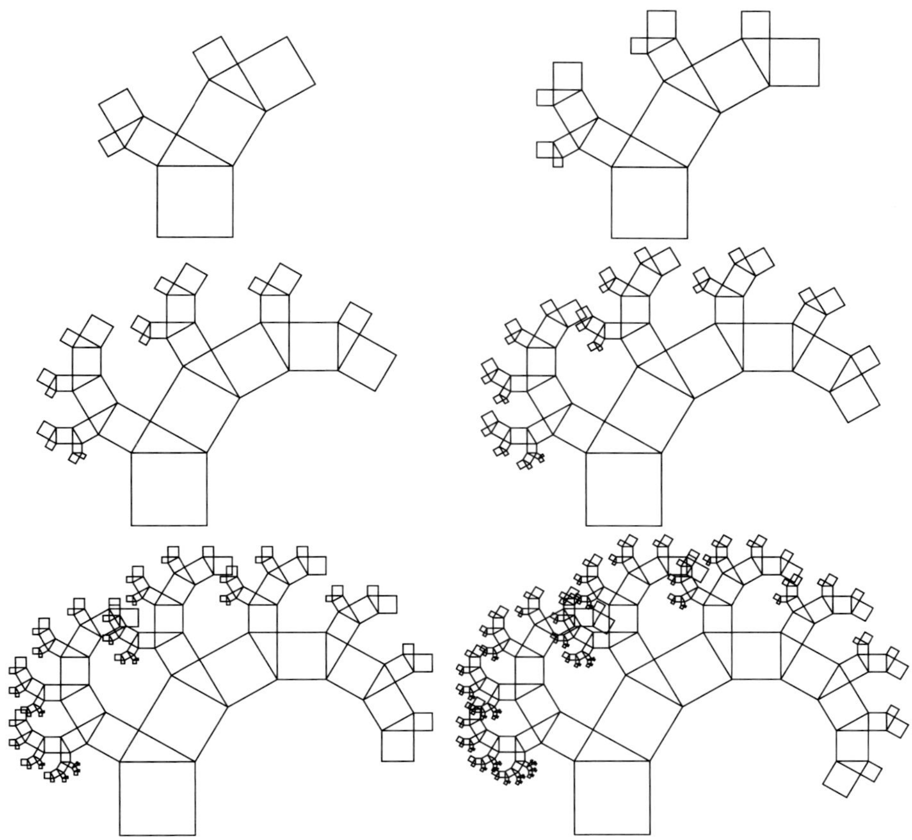

FIGURE 11. More iterations of the 30-60-90 Pythagorean Tree Fractal. The global structure starts to become apparent, and differs from the isosceles version.

Indeed, this puts us precisely in the setup of Definition 2.1. For angles θ in the interval $[30, 45]$, we can let $P_\theta \subseteq \mathbb{R}^2$ be the Pythagorean tree fractal constructed using the right triangle of minimal angle θ in the recursive step, considered as a

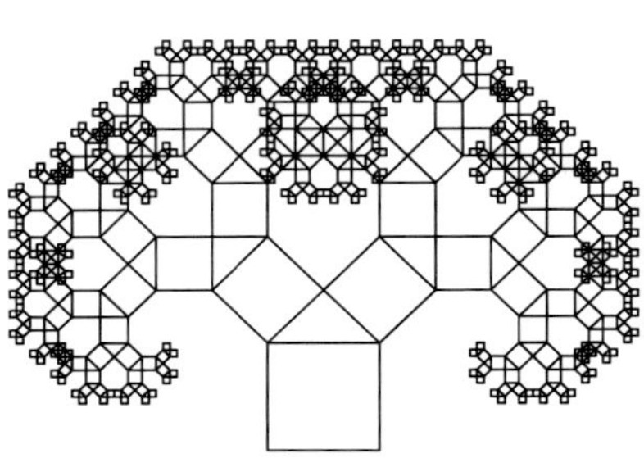

$\theta = 45$

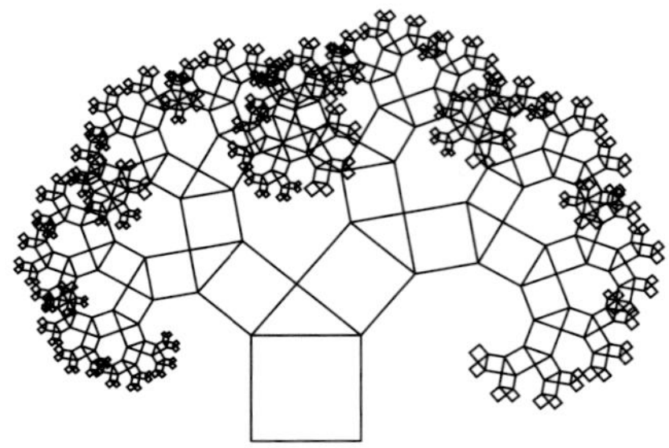

$\theta = 40$

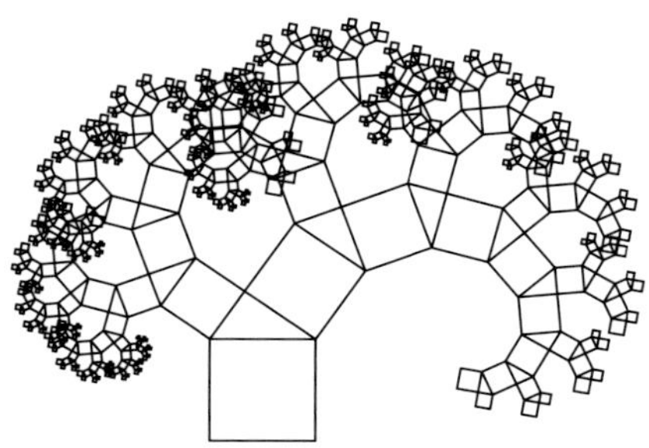

$\theta = 35$

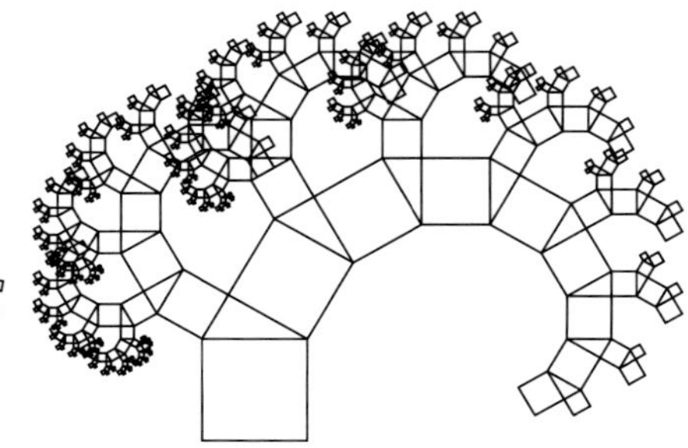

$\theta = 30$

FIGURE 12. Pythagorean tree fractals for various types of right triangles.

filled subset of the plane. This gives us an entire deforming family of Pythagorean tree fractals, which one can consider as an animation on a screen [**3**].[6]

3D printing gives us a new way to observe the transformation of this fractal as θ varies. Indeed, our approach is to follow the framework of Section 2, and to consider the deformation space as a solid static animation in $\mathbb{R}^3$:

$$\mathcal{P} = \{(x, \theta) : \theta \in [30, 45], x \in P_\theta\} \subseteq \mathbb{R}^3.$$

We can use a computer to model $\mathcal{P}$ using CAD software (we did this using the OpenSCAD [**16**]), and create a 3D object which can be 3D printed, held in your hand, and studied from all points of view (Figure 13). In this way, 3D printing gives us a new avenue to illustrate perturbations and deformations of this family of fractals. For one, we get a new dimension where we can observe the movement−for example, by considering where the slope becomes more exaggerated, it becomes immediately visible that the change in angle has more of an effect toward the outer edges of the fractal, and more of an effect on the *30 degree* side than the *60 degree side*. Even more, one obtains a physical and tactile way to experience a mathematical concept, opening new doors for broad mathematical interaction.

[6]See Matthew Conroy's animation as θ varies from 0 to 90: `https://www.madandmoonly.com/doctormatt/webimages/animations/pythagoreanFractal2.gif`.

FIGURE 13. A 3D printed static animation depicting the deformation of the Pythagorean tree Fractal. Photo credit: Edmund Harriss, University of Arkansas.

4. Deformations of Julia sets

Joint with Bernat Espigulé. Let $\mathbb{C}$ denote the field of complex numbers, fix a complex number $c \in \mathbb{C}$, and consider the function $f_c : \mathbb{C} \to \mathbb{C}$ given by the rule $f_c(z) = z^2 + c$. What happens when we iterate this function? More explicitly, if we fix some $z_0 \in \mathbb{C}$, what can we say about the following sequence of complex numbers?

$$\begin{aligned} &z_0 \\ &z_1 = f_c(z_0) \\ &z_2 = f_c(z_1) = f_c(f_c(z_0)) \\ &z_3 = f_c(z_2) = f_c(f_c(f_c(z_0))) \\ &\vdots \end{aligned}$$

A fundamental question asks how the behavior of this sequence changes as we vary z_0, or as we vary c. Let's get a basic idea of what can happen by letting $c = 0$.

EXAMPLE 4.1. Consider $f_0(z) = z^2$. Then

$$f_0^n(z_0) = \underbrace{f_0 \circ f_0 \circ \cdots \circ f_0}_{n-\text{times}}(z_0) = z_0^{2n}.$$

If $|z_0| < 1$ then this sequence converges to 0, if $|z_0| > 1$ then this sequence escapes to infinity, and if $|z_0| = 1$ then this sequence stays on unit circle $|z| = 1$. In particular, the set

$$\{z, f_0(z), f_0^2(z), f_0^3(z), \cdots\}$$

is bounded if and only if $|z| \leq 1$.

It appears that when iterating f_c from a fixed starting point we can expect the outputs to either escape to infinity, or else to all be contained in a bounded region of $\mathbb{C}$. Filled Julia sets consist of the starting points which produce bounded sequences.

DEFINITION 4.2. Let $c \in \mathbb{C}$ be a complex number and consider the function $f_c(z) = z^2 + c$. The *filled Julia set* associated to c is the set:

$$J_c := \{z \in \mathbb{C} : \text{ the set } \{z, f_c(z), f_c^2(z), \dots\} \text{ is bounded}\} \subseteq \mathbb{C}.$$

In Example 4.1 we determined that the filled Julia set J_0 is precisely the complex unit disk $\{z : |z| \leq 1\} \subseteq \mathbb{C}$. For other complex numbers, the situation becomes much richer. For example, in Figure 14 are plots of J_c for various values of c[7].

A central question asks how J_c changes as c does. For example, in Figure 15 we observe the variation of the filled in Julia sets for polynomials f_c with values $c = -.75$, $c = -.74 + .05i$, and $c = -.73 + .1i$ respectively. It appears that if c changes a little bit, then so does J_c,[8] suggesting that we can use these filled Julia sets as the frames of a static animation. But, as discussed in Section 2.2, there is an obstacle. This is because the varying parameter c is a complex number, and so we obtain a 2-dimensional stack of frames. When we form a deformation space exactly as in Definition 2.1, we obtain a subset of $\mathbb{C} \times \mathbb{C}$:

$$\mathcal{J} = \{(c, z) : c \in \mathbb{C}, z \in J_c\} \subseteq \mathbb{C}^2.$$

As a hypersurface in 4D space, this object is impossible to 3D print. In order to get something we *can* print, we will choose a 1-dimensional path through the parameter space (c.f. Section 2.2). But which paths should we choose? To answer this it is important to understand a little more about the geometry of the parameter space of filled Julia sets.

Each filled Julia set we have looked at so far has been connected and has a clearly defined interior, but for many values of c the set J_c is a totally disconnected Cantor set [**17**, Chapter 9]. For the purposes of 3D printing it makes sense to focus on values c where J_c is connected, which is a fascinating and celebrated collection of complex numbers in its own right.

DEFINITION 4.3. The *Mandelbrot set* is the set:

$$\mathcal{M} := \{c \in \mathbb{C} : J_c \text{ is connected}\} \subseteq \mathbb{C}$$

[7]The 2D images of filled Julia sets in this section were rendered in Mathematica [**26**]

[8]In fact, the situation is quite a bit more subtle than this. We discuss this a little further in Section 4.2. The question is studied systematically by Douady [**9**].

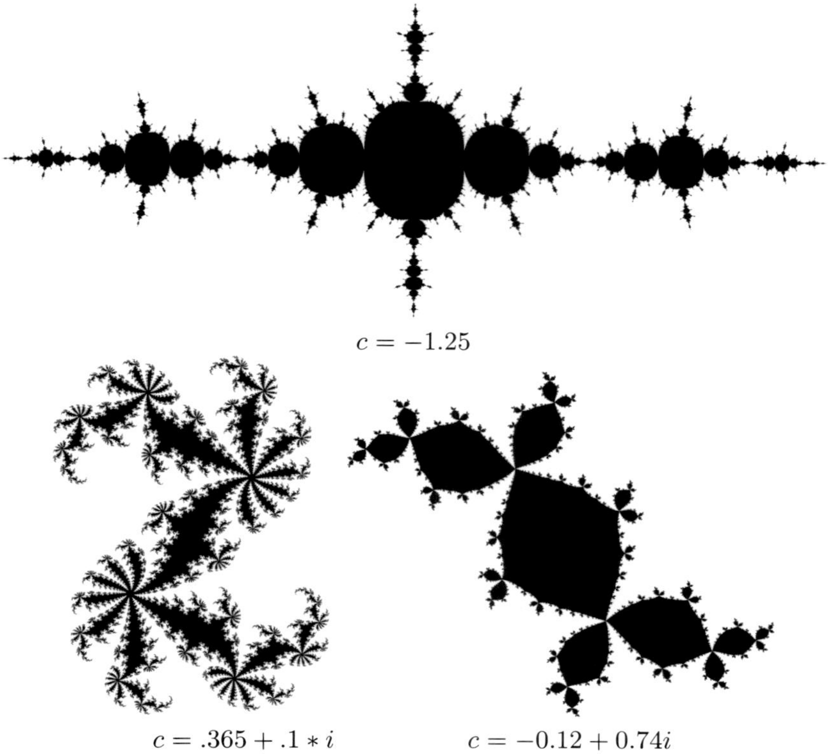

FIGURE 14. Julia sets for various complex parameters.

A more familiar definition of the Mandelbrot set is perhaps

$$\{c \in \mathbb{C} : \{0, f_c(0), f_c^2(0), \dots\} \text{ is bounded}\} = \{c \in \mathbb{C} : 0 \in J_c\}.$$

The fact that these are equivalent is [**17**, Theorem 9.5, Appendix G]. The Mandelbrot set has a famous shape, depicted in Figure 16. To cut down our deformation space $\mathcal{J}$ by one dimension and obtain something 3D printable, we mirror Section 2.2 and pick a 1-dimensional subset of the Mandelbrot set defined by a path. As the behavior of the Julia set J_c is closely related to the location of c in the Mandelbrot set, we will use the geometry of the Mandelbrot set to inform our choice of path.

4.1. Turning paths into prints. Let $I \subseteq \mathbb{R}$ be an interval, and consider a continuous map

$$\gamma : I \to \mathbb{C},$$

whose image is contained in $\mathcal{M}$. For each $t \in I$, we can consider the filled Julia set $J_{\gamma(t)}$. This gives us a family of filled Julia sets parametrized by t, putting us squarely in the situation of Definition 2.1. Therefore the deformation space associated to this data is a static animation:

$$\mathcal{J}_\gamma := \{(z, t) : t \in I, z \in J_{\gamma(t)}\} \subseteq \mathbb{C} \times \mathbb{R} \approx \mathbb{R}^3.$$

This sets up a general framework for creating static animations illustrating the deformation of filled Julia sets associated to a path in the Mandelbrot set. It only

FIGURE 15. Julia sets associated to nearby complex parameters.

remains to choose a path. The boundary of the Mandelbrot set partitions the collection of filled Julia sets into two types: the filled Julia sets that are connected, and those that are not. In particular, it behaves like a critical point, and we begin by studying how filled Julia sets deform near the boundary, *or along it.* Our initial exploration was to traverse the upper boundary of the main cardioid, depicted in red in Figure 17. One can deduce from [**17**, Chapter 10] that the boundary of the

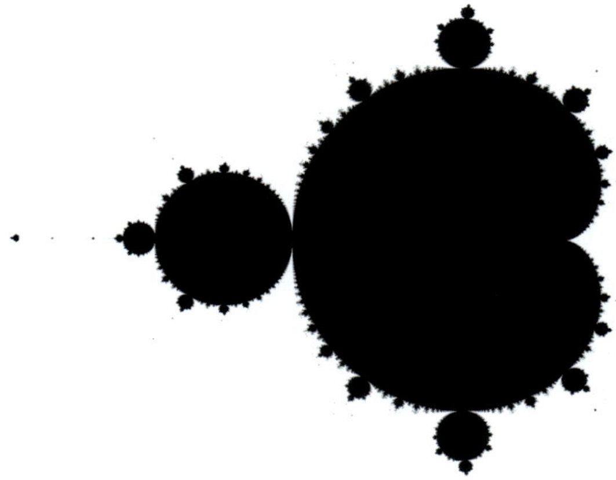

FIGURE 16. The Mandelbrot Set.

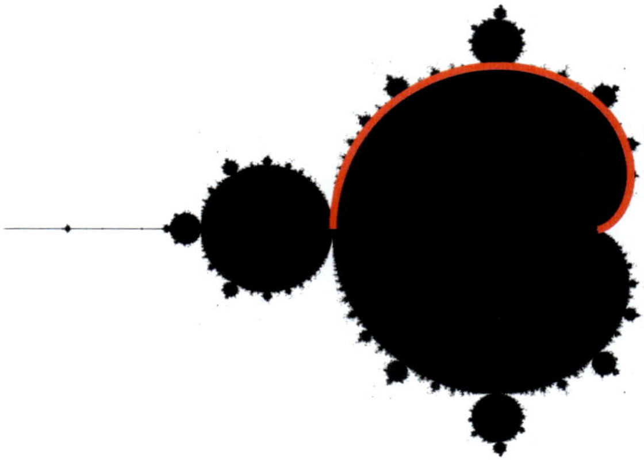

FIGURE 17. The Mandelbrot Set, with the path along the top of the main cardioid drawn in red.

main cardioid can be parametrized as follows.

$$x(t) = \frac{1}{2}\cos(t)(1-\cos(t)) + \frac{1}{4},$$
$$y(t) = \frac{1}{2}\sin(t)(1-\cos(t)).$$

Then the path $\gamma : [0,\pi] \to \mathbb{C}$ given by $\gamma(t) = x(t) + iy(t)$ parametrizes the upper boundary of the main cardioid. Associated to this path there is a static aimation $J_\gamma \subseteq \mathbb{C} \times \mathbb{R}$, a 3D print of which is pictured in Figure 18.

4.2. Parabolic implosions. We suggested in Section 2 that we were mostly interested in static animations with continuously deforming frames, but is this true for Julia sets? This question was studied by Douady [**9**] who showed that in general, it is not. The fact is, there are some special points at which $c \mapsto J_c$ does not vary continuously in c, even along our path γ traversing the main cardioid. To illustrate an example of this we look at the filled Julia sets associated to $\gamma(0)$ and $\gamma(\varepsilon)$ for

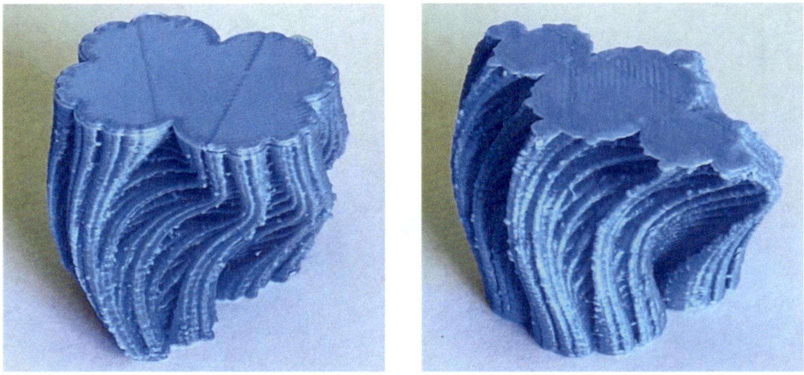

FIGURE 18. 3D printed static animations associated to a traverse of the main cardioid.

FIGURE 19. Two Julia sets suggesting a failure of continuity of the deformation.

some very small ε (in this case $\varepsilon = .075$). In Figure 19 we show the frames J_c associated with the complex numbers $c_0 = .25$ and $c_\varepsilon = 0.251402 + 0.000105321i$ respectively. Although the overall shape is similar, there is a sudden appearance of a number of gaps. This is an example of a phenomenon called *parabolic implosion* [**9**, Section II], which is of considerable interest [**20**], but this curious phenomenon isn't clearly visible in our model, as the filled top layer of the print obstructs the view of the suddenly appearing gaps. One way to see the parabolic implosions is to instead consider a restriction of our path to a slightly smaller interval. There is a parabolic implosion also at $\gamma(\pi)$, so we consider the path $\gamma : [\varepsilon, \pi - \varepsilon]$ for very small ε. The static animation associated to this path allows a better view of formation and deformation of the parabolic implosions, and produces a striking visual effect (Figure 20). Having a 2-dimensional parameter space initially seemed like an obstacle in creating static animations, but the philosophy of Section 2.2 turns this into a richness instead. The freedom of having a static animation associated to each path allows us to focus on interesting areas and study them from multiple points of view. We exploited this by choosing paths that remain in the vicinity of these parabolic points, allowing us to do local explorations around the parabolic implosions of the cusp associated with $c = \frac{1}{4}$, by considering nearby values of c, constructing deformed Julia sets for a continuously varying value $c \approx \frac{1}{4}$ (Figure 21). We do a similar exploration around the parabolic implosion near the so-called

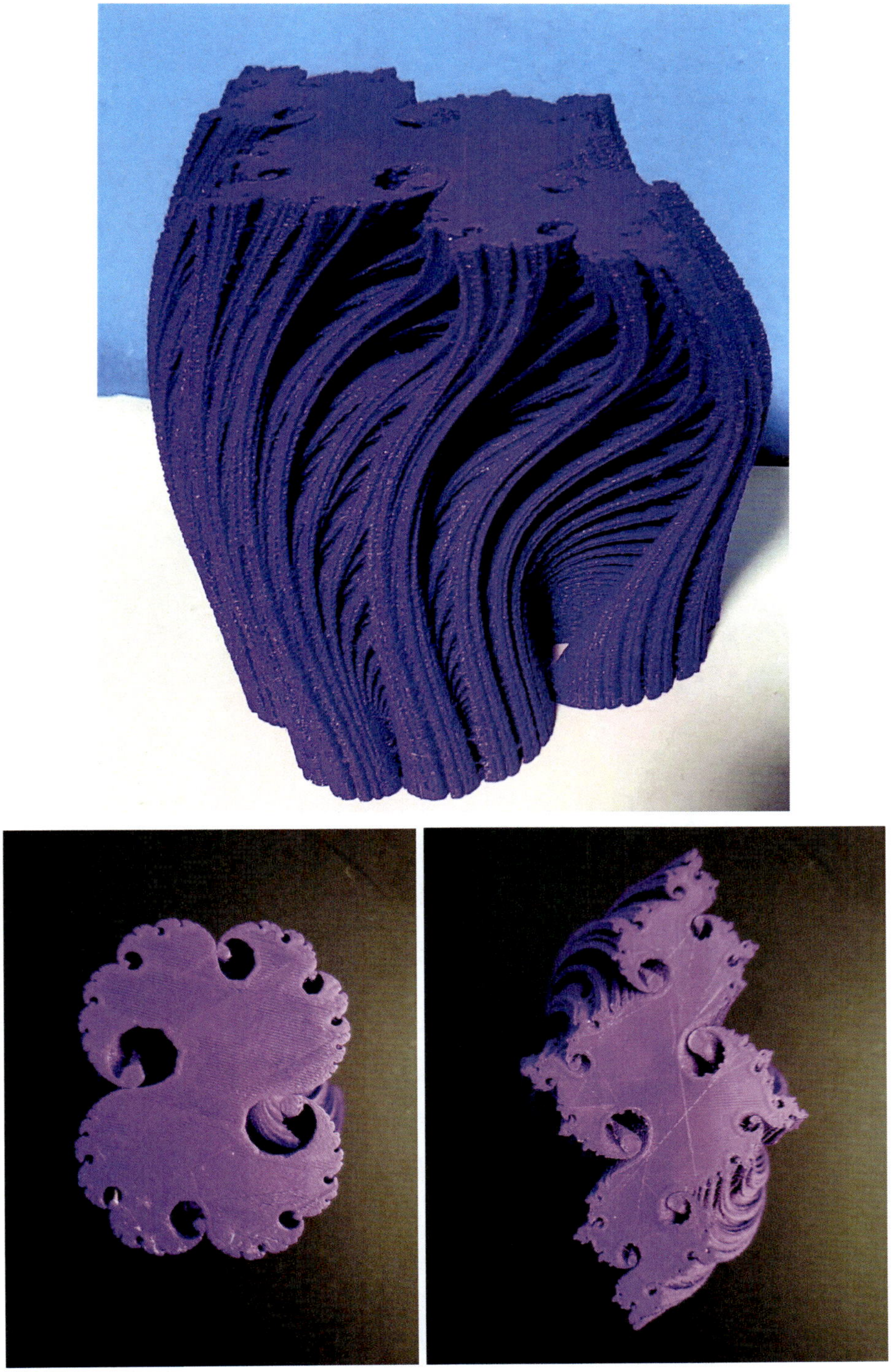

FIGURE 20. 3 views of the 3D printed deformation space associated to a traverse of the main cardioid, leaving the parabolic implosions visible.

FIGURE 21. 3D printed static animation for complex parameters near the cusp: $c \approx \frac{1}{4}$. Photo credit: Edmund Harriss, University of Arkansas.

FIGURE 22. 3D printed static animation for complex parameters near the Basilica: $c \approx -\frac{3}{4}$. Photo credit: Edmund Harriss, University of Arkansas.

Basilica where $c = -\frac{3}{4}$, by again varying our complex input c to nearby values and constructing the deformed Julia sets for a continuously varying value $c \approx -\frac{3}{4}$ (Figure 22).

4.3. Non-connected Julia sets. So far we have restricted ourselves to considering filled Julia sets J_c for values of c in the Mandelbrot set. What happens if we try to also consider J_c for $c \in \mathbb{C} \setminus \mathcal{M}$? These Julia sets will not be connected, and in fact, the Julia-Fatou theorem says that J_c is either connected, or else totally disconnected and homeomorphic to a Cantor set [**17**, Theorem 9.5]. We experimented with how our methods (cf. Section 4.1) adapted to a path which exited the Mandelbrot set, and obtained a curious result (Figure 23). Where do these towering mountains come from? In some sense, they are an artifact of computer graphics. What happens is that our method for generating images of Julia sets uses density to color pixels. That is, if a pixel contains many points in the Julia set, it will be colored black, and otherwise it will be left white. As a result, for complex numbers c which are very close to the boundary of $\mathcal{M}$, the screen displays J_c as a union of several connected components, since these are where there are high

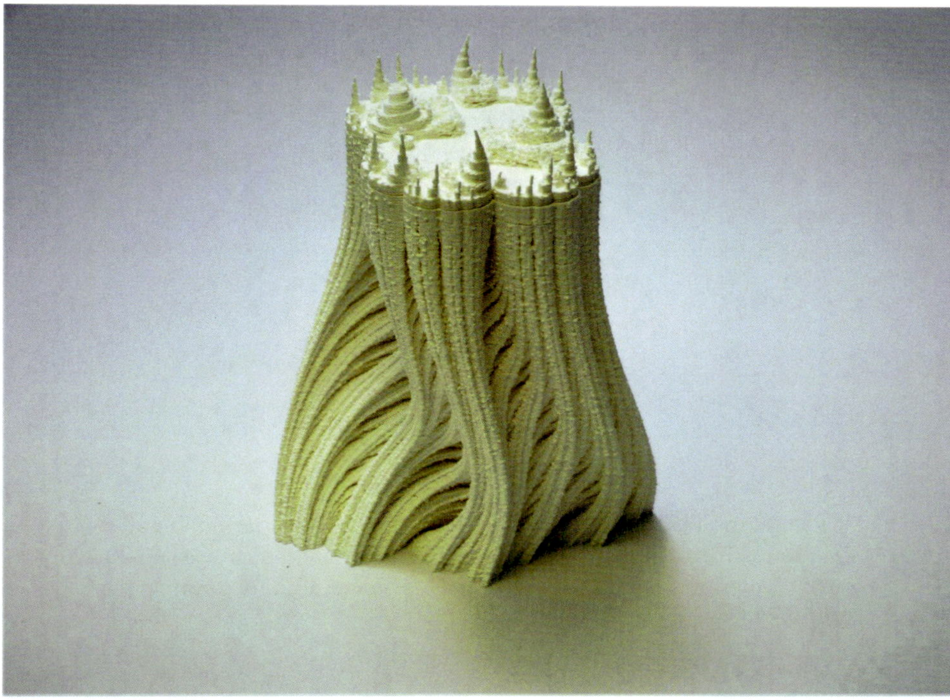

FIGURE 23. 3D printed static animation associated to a path that leaves the Mandelbrot set.

densities of points in the Julia set. For example, Figure 24 shows our image of the Julia set for for $c = .28 + .01i$, a complex number outside of the Mandelbrot set, but just barely. *We emphasize that this is an artifact of computer graphics*, the set J_c is totally disconnected. What appears to be a connected region is merely a part of the plane where there is a high concentration of points which are bounded under the dynamical system of iterating f_c. As c progresses further away from $\mathcal{M}$, these high concentration areas spread out and we can observe a rapid progression toward a more sparse Julia set. It is worth remarking that this print, although not mathematically rigorous in the purest sense, provides mathematical insights about concentrations of points in Cantor-like Julia sets and how they deform and spread out.

The fact that an artifact of computer graphics can both create interesting aesthetics and contribute meaningfully to our understanding is interesting in its own right, although beyond the scope of this article. Discussions about techniques, artifacts, and limitations of generating Julia sets using computer graphics are discussed by Milnor [**17**, Appendix H], and in great depth by Chéritat [**2**].

4.4. Other directions. The static animations of filled Julia sets exhibited so far have been limited to paths on and near the boundary of the main cardioid, but there is an entire menagerie of static animations that can be constructed simply by choosing paths through the Mandelbrot set. We invite the reader to explore these possibilities, and to this end we have included a Mathematica notebook called

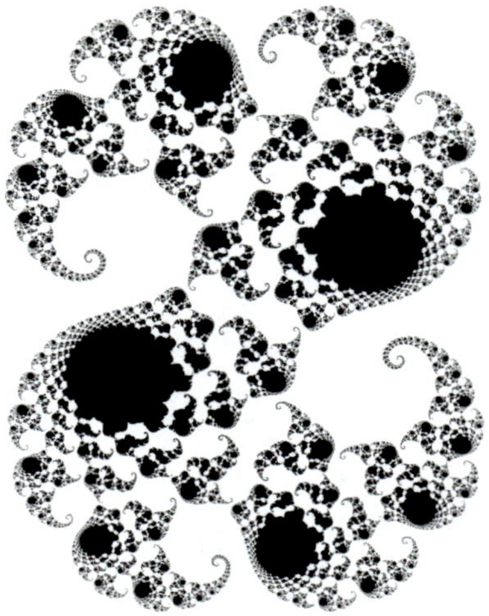

FIGURE 24. The Julia set associated to complex parameter $c = .28 + .01i$. In reality, this set is totally disconnected, but due to a quirk of computer graphics in how the image is made, connected components seem to appear.

FIGURE 25. A computer generated solid depicting the deformation space associated to a traverse of the boundary of the period 2 hyperbolic component of the Mandelbrot set.

`JuliaPlots.nb` in the public github repository accompanying this paper [**5**]. This notebook allows the user to specify a parametrized path through $\mathbb{C}$, and will generate the frames of the associated static animation. The reader can then follow the procedure described in Section 5 to construct the resulting static animation. For now, we suggest a few of the many paths worth exploring, and highlight a related project led by Caroline Davis [**1**].

4.4.1. *Traversing other boundaries.* To get a more complete picture of how filled Julia sets deform along the boundary of the Mandelbrot set we should also consider static animations associated to the boundaries of other bulbs. For example, the bulb directly left of the main cardioid (called the *period 2 hyperbolic component*) is a perfect circle of radius $1/4$ centered at -1. Figure 25 is the static animation associated to traversing the upper semicircle of this component. The other hyperbolic components do not have boundaries that are so easy to parametrize. Giarruso and Fisher explicitly parametrized the period 3 hyperbolic component [**11**], and the parametrization is much more complicated than the two we have seen so far.

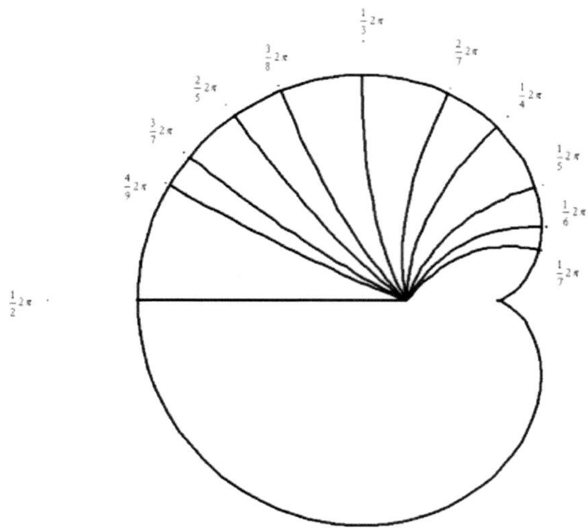

FIGURE 26. Various radii of the main cardioid of the Mandelbrot set[**12**]. Reproduced with permission from Walter Hannah.

Beyond that, little has been worked out explicitly, leaving the question of how to correctly traverse the boundary as an interesting and difficult one. In fact, the question of whether the boundary of the Mandelbrot set can be parametrized as a curve at all is related to the famous Mandelbrot Locally Connected (MLC) conjecture of Douady and Hubbard [**8**],[**7**]. It probably isn't quite as difficult to find the boundary of a specific hyperbolic component, nevertheless, the problem of finding parametrizations of subsets of the boundary of the Mandelbrot set is connected to deep research questions in complex dynamics.

4.4.2. *Traversing radii.* The boundary is not the only interesting place to look. Another direction (suggested by Caroline Davis), would be to consider the *radii* of the hyperbolic components. Indeed, each bulb of the Mandelbrot Set gives rise to Julia sets with different discrete invariants (such as the periods of the *attracting cycles* [**17**, Chapter 12]). It is also known that each component is canonically holomorphic to a circle [**18**], so that each bulb has a canonically defined center and radii. Tracing the radii between the center of one bulb to the center of another gives rise to a canonical deformation between two Julia sets with distinct discrete invariants, something that we are interested in 3D printing. An important step is to obtain explicit parametrizations of these radii. As with the boundaries, these are easy to find for the main cardioid, and the hyperbolic component of period 2 (which is already a disk), and can be extracted from [**11**] in the period 3 case. Figure 26 is an image made by Walter Hannah [**12**] depicting these radii in the main cardioid.

4.4.3. *Related objects in holomorphic dynamics.* We have only considered Julia sets associated to quadratic polynomials, but one can also iterate other functions–such as polynomials of higher degree or rational functions–and study their associated Julia sets. An undergraduate research group at Indiana University, led by Caroline Davis, studied a family of polynomials whose associated Julia sets interpolate between the Apolonian and Sierpinski gaskets [**1**]. Figure 27 shows a print of the associated static animation, printed by Bethany Mussman.

FIGURE 27. A static animation of a Julia set associated to a higher degree polynomial[**1**]. Print by Bethany Mussman of a design by Caroline Davis.

4.5. Static vs. computer animations of Julia sets. We are certainly not the first to visualize how Julia sets deform according to carefully chosen paths in the Mandelbrot set. There are many computer animated visualizations of deforming Julia sets, including witnessing degenerations from connected Julia sets to Cantor sets, and exploratations near parabolic implosions.[9] The shift from computer to static animations brings with it advantages and disadvantages. For example, computer animations give a more direct view of the internal structure, which can be hidden by the walls of the static animation, although this can be mostly remedied by adjusting the start and endpoints of the chosen path like we did in Section 4.2. Nevertheless, there is a completely novel perspective the static animation brings to the table, something that is missing from the computer animation: *the side view.* From the horizontal perspective we can observe how the different bulbs on the edge of these Julia sets move at different rates as the Julia set deforms, and these bulbs braid around each other in a way that seems to interpolate between the mathematical and the organic. Figure 28 gives a particularly good illustration of this phenomenon.

5. Creating a static animation

When entering the world of 3D printing, it is easy to get overwhelmed by the large number of CAD programs and 3D modeling tools, many of which can be difficult to learn, especially as a beginner. On the other hand, since static animations consist of stacks of 2-dimensional objects, one can hope to build a 3D static animation using purely 2-dimensional techniques. In this section we describe

[9]In fact, the suggestion to 3D print parabolic implosions was made to the author by John H. Hubbard as he was demonstrating an animation he made of this phenomenon.

FIGURE 28. A side view of the static animation associated to the traverse of the main cardioid. This perspective gives a glimpse which seems to interpolate between the mathematical and the organic.

a workflow that does exactly that,[10] so that anyone who can make a 2-dimensional computer animation can begin designing 3D prints. After describing the general process, we will then run through an explicit tutorial, building the static animation of deforming polar flowers from Section 2. We intend this section to serve as an invitation for beginners to mathematical 3D printing, while also providing a new modeling technique to the experts.

The main idea is simple, and can be thought of as a sort of *reverse tomography*. While tomography is an imaging technique which extracts a collection of 2D cross sections from a 3D object, we begin with a collection of 2-dimensional frames, which

[10]This particular workflow was first shared with the author by Bernat Espigulé, and is how we created the static animations of filled Julia sets described in Section 4.

we stitch together into a 3D model whose horizontal cross sections will be precisely the frames we begin with.

A summary of the workflow is the following: create a collection of 2-dimensional frames however you like, and upload them to a piece of software that can bond them together into a 3D model. We use the open source software, UCSF Chimera [**21**],[**25**]. It has many uses, but we are most interested in its use as medical imaging software. Roughly, MRI tomography takes a series of images of an organ at various depths inside the body, and these images can be fed to Chimera to reverse engineer a 3D model of the organ. We can exploit this functionality to make our static animations.

5.1. Overview. Fix a collection of frames B_t, parametrized by values t in some interval I. We will describe a general process that can be followed to model the static animation

$$\mathcal{B} = \{(b,t) : b \in B_t\} \subseteq \mathbb{R}^3,$$

as a 3D object file. These steps assume the reader can already produce a 2-dimensional image of each frame B_t.

Step 1: Make a collection of frames. Subdivide your interval into a reasonable number of frames. Keep in mind that a common layer thickness for a 3D print is about .2mm, and the larger at home 3D printers allow a maximum height of no more than 300mm, so the resolution of your 3D printed object will probably not benefit from having more than 1500 frames (although industrial 3D printers can do better). This gives a finite number of $t_i \in I$ which you will sample from.

For each t_i, produce a **PNG image** of B_{t_i}, coloring points $p \in B_{t_i}$ white, and points $q \notin B_{t_i}$ black. This is the color scheme and file type that Chimera recognizes as cross sections of a 3D object. Save the entire collection of images in a single folder and choose filenames so that the alphabetical ordering of the images corresponds to the correct ordering of the frames, this way they will be imported in the correct order during the next step.

You can create the frames of the animation any way you would like. For the prints in Section 4, we fed a sampling of points on our parametrized path to Mathematica's JuliaSetPlot function [**26**], which we modified to color points in white and black as needed.

Warning: You may need to remove the alpha channel. The PNG graphics file format can store color images with an RGB color palette (storing numerical values for the red, green, and blue of each pixel), or an RGBA color palette, which has a fourth *alpha* channel for transparency. *Chimera will reject PNG graphics with an alpha channel*, so you may need to manually remove the alpha channel from your frames. To do this we wrote a short script called `alphaRemover.py`, included in the repository accompanying this paper [**5**].

Step 2: Convert the frames to a solid in Chimera. From the Chimera home screen, choose `File-->Open`. Navigate to and highlight the entire set of images from Step 1, and select `Open`. A rough version of your model should become visible, as well as a *volume viewer* menu with various settings. There are 3 important settings to be aware of:

- **Step:** A dropdown menu labelled *step*, with choices of values 1,2,4,8, and 16 (4 is the default). This roughly controls the vertical resolution. *Choose*

1 (the highest vertical resolution). Your model should immediately look much sharper.

- **Level:** A slider to select the *level*, which is a number between 0 and 255. This roughly controls the tolerance for edge detection. We often have to play around with this a bit.
- **Style:** Radio buttons to select either surface, mesh, or solid. We always choose *surface*.

Once you have found settings to your liking, you can export the model. Select `File-->Export Scene...` You can choose from a number of file types. The standards for 3D printing are OBJ or STL. Depending on the size of the image and number of frames, *this can take a long time.* For the particularly detailed animations of Section 4 we needed to let it run for several hours.[11]

Step 3: Slice and print. Your model should now be ready to open up in a slicer (for example *Cura* [**4**]), and then sliced and sent to a 3D printer. The full process depends on your 3D printer, and is beyond the scope of this article, but we'd like to include a few tips to help avoid failure at this step. We point to a few resources for slicing and printing 3D models at the end of this subsection.

Avoid steep slopes and overhangs. An extrusion based 3D printer will print your static animation by printing each frame on top of the last, using the previously printed frames as a support base for the next ones. If the frames B_t vary too wildly with t, the printer may end up trying to extrude part of a frame with no supporting structure underneath. These unsupported layers of plastic building up will lead to unsatisfactory results and failed prints. As a rule of thumb, most 3D printers can cleanly print overhanging structures with an angle up to somewhere between 45 and 60 degrees, so if you would like to print a wildly deforming family, you may need to do some experimentation to get something that works.[12]

Lay your model flat. Chimera won't automatically export a model which lies flat, instead exporting the model aligned to the perspective in the viewing window. Sending the tilted model to an extrusion based 3D printer will likely cause issues with adhesion and support, and could result in a failed print. This can be avoided if you don't adjust the view of the model in Chimera's viewscreen between importing your frames and exporting the STL. Otherwise you must rotate your exported model so that it lies flat on the buildplate. This can be done in Cura using the *Lay Flat* command.

More resources for the printing process. Although beyond the scope of this paper, there are many places to turn for a thorough introduction to the use of slicers and 3D printers to turn a digital model like your static animation into a physical model. For example, the author's lecture during the 2022 Joint Mathematics Meetings mini-course *3D Printing: Challenges and Applications* was a tutorial on slicing and printing, and the slides are publicly available [**6**]. There are also many detailed written tutorials (for example [**10**]) and youtube tutorials (for example [**19**]).

5.2. Tutorial. We now demonstrate this process step-by-step with an example, constructing the static animation of deforming polar flowers introduced in Section 2. The reader is invited to follow along on their own machine. To this

[11]For reference, the computation was done on a modestly powered Dell XPS 15 7590 with an Intel Core i7-9750H (12MB Cache, up to 4.5 GHz, 6 cores).

[12]Alternatively, you can employ the use of *support material*, extra printed material which holds up excessive overhangs and can be removed after.

FIGURE 29. A filled polar star. We fill the star to make our static animations solid.

end we have provided code snippets within this document, as well as the complete collection of code and files in a public Github repository [**5**]. For more detail on where to find what you need to follow along, see Section 5.3 below.

REMARK 5.1. In Section 2.1 we computed a complete parametrization of this particular deformation space in $\mathbb{R}^3$. With this parametrization there are many more direct ways to produce a 3D model (for example, plotting in Mathematica or Sagemath and then exporting as an STL). Nevertheless, one often encounters deformation spaces which do not admit such a parametrization (for example, those in Sections 3 and 4). One should think of this tutorial as an introduction to this particular process, rather than the most efficient way to make this particular model.

The frames of the static animation in this tutorial are polar curves whose equations we computed in Section 2.1 to be $r = 2 + t\cos(5\theta + 2\pi t)$ for values t in the interval $I = [0.2, 1]$. Since we are making a 3-dimensional solid, we will replace these curves with the regions they bound, so that our resulting object will be filled rather than hollow. The frames we will use will therefore be the following sets

$$B_t = \{(r, \theta) : r \leq 2 + t\cos(5\theta + 2\pi t)\} \subseteq \mathbb{R}^2.$$

For example, B_1 is now the entire shaded region in Figure 29, rather than just the boundary curve. We now run through all the steps to model the static animation $\mathcal{B} = \{(b, t) : t \in I, b \in B_t\}$ as a 3D object file.

(Tutorial) Step 1. We subdivide our interval to create 161 frames, so that t progresses between 0.2 and 1 at steps of .005 units. Our list of t values is therefore

$$T = \{.2, .205, .21, .215, \cdots, .990, .995, 1\}.$$

We need to make a PNG image for each frame, drawing each region B_t in white, against a black background. Let's walk through a way to do this in python. We need the `matplotlib` and `numpy` libraries.

```
import matplotlib.pyplot as plt
import numpy as np
```

Chimera expects white images against a black background, so we first set the color of the background we are drawing against.

```
plt.style.use('dark_background')
```

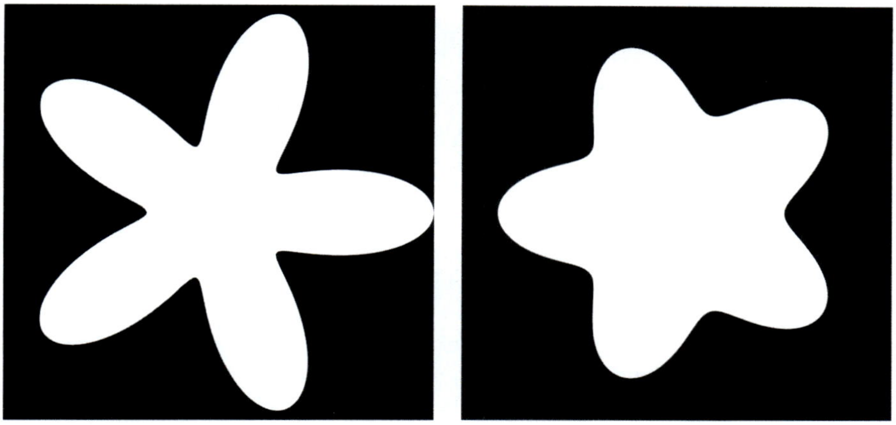

FIGURE 30. 2 polar star. We draw the insides white so that they are compatible with Chimera.

For each frame, we'd like to draw the polar curve $r = 2t\cos(\theta + 2\pi t)$. To do this, `mathplotlib` will sample a number of points and interpolate. We use numpy's `linspace` command to create the list of θ values to sample from.

```
NUM_PTS = 1000
theta = np.linspace(0,2*np.pi,NUM_POINTS)
```

We then calculate the corresponding r values to finish computing the list of points. Notice we set $t = 1$ in this snippet, but this will change from frame to frame.

```
t = 1 #This varies on each frame
r = 2 + t*np.cos(5*theta + 2*np.pi*t)
```

Now whats left is to let `mathplotlib` draw and fill the curve.

```
plt.polar(theta,r,alpha=0) #Draws the curve
plt.fill_between(theta,0,r,color='white') #Fills the curve
plt.axis('off') #We just want the flower, no axes
```

Figure 30 displays is the output with t values 1 and 1/2 respectively.

To create the entire stack of frames we loop through each value of $t \in T$ and export a PNG built using the the code above, choosing filenames in alphabetical order so that they are arranged correctly when imported by Chimera. We can do this by looping through the last two snippets of the code above and saving the output for each frame.

```
NUM_FRAMES = 161 #Set the number of frames

for i in range(NUM_FRAMES):
  t = 0.2 + .005*i
  r = 2 + t*np.cos(5*theta + 2*np.pi*t) #The boundary curve
  plt.polar(theta,r,alpha=0) #Draws the curve
  plt.fill_between(theta,0,r,color='white') #Fills the curve
  plt.axis('off') #We just want the flower, no axes

  #Now export and clear the plot.
  imageName = 'flowerStack/flowers_'+str(1000+i)+'.png'
```

```
    plt.savefig(imageName) #Export the image
    plt.cla() #clear the plot before looping back
```

The images are now saved as `flowers_1000.png` through `flowers_1160.png`. The reason we started counting at 1000 is because Chimera reads alphabetically in such a way that 10 comes before 9 because it starts with a 1. In particular, if we started counting at 0 Chimera would arrange the imported frames in an incorrect order during Step 2. We now have our frames images all saved as PNG images (Figure 31). In our accompanying github repository [**5**] we include a python program `flowerPlot.py` for creating the entire stack of frames, as well as an implementation in Mathematica, `flowerPlot.nb`.

(Tutorial) Step 2. Open Chimera, select browse, and highlight all of the PNG images from step 1 (Figure 32). Click open. Chimera stitches all of the polar flowers into a single solid (Figure 33). It looks a bit rough. By changing *step* to 1 and adjusting the level we arrive at a much finer model. We also choose *surface* in the style setting if it isn't already selected (Figure 34). This looks much better.

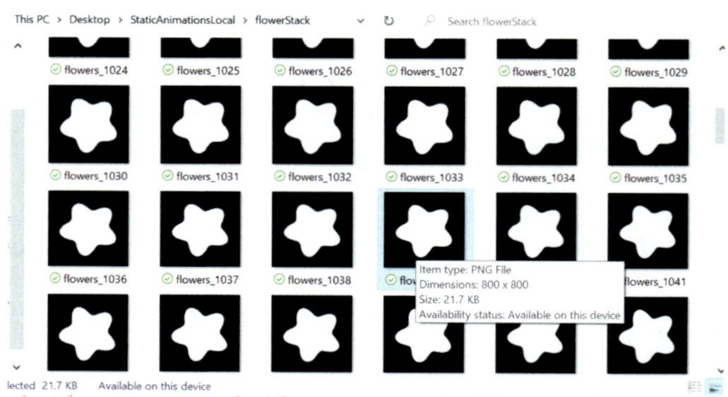

FIGURE 31. A folder containing the collection of deforming polar flowers.

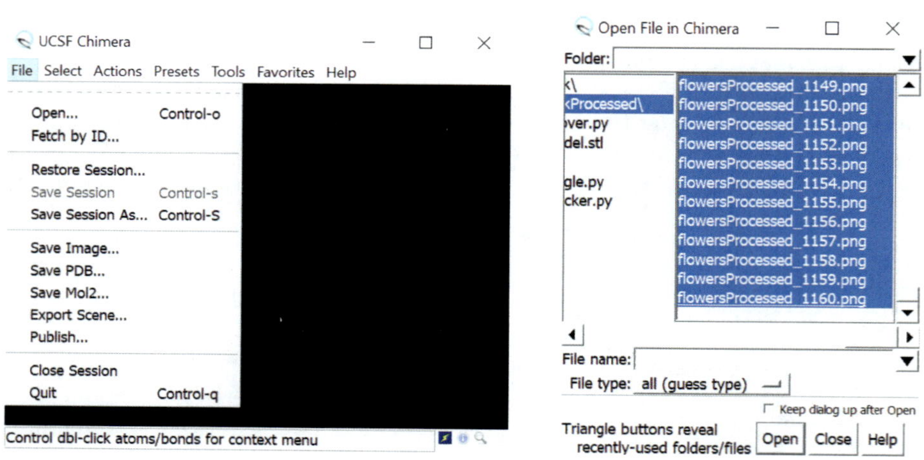

FIGURE 32. The process for importing the images to Chimera.

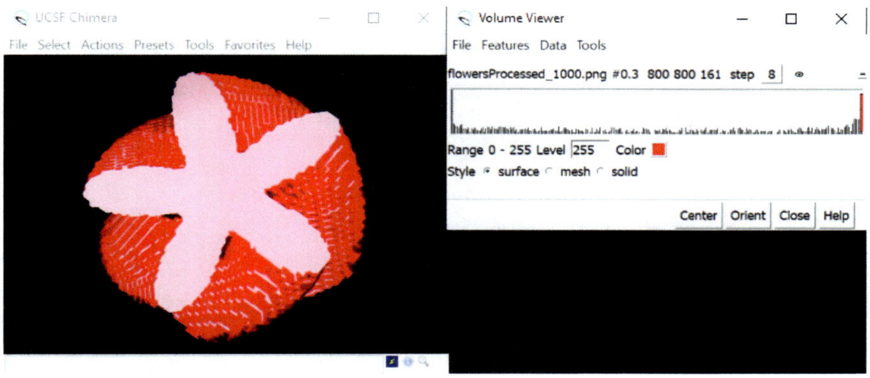

FIGURE 33. What appears on the screen after importing our image stack.

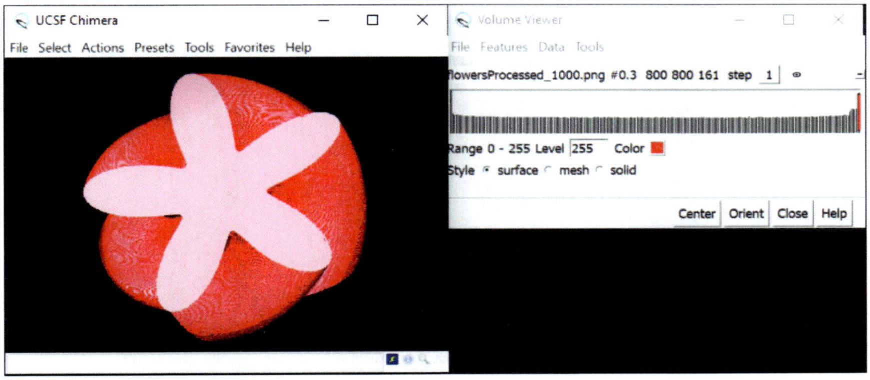

FIGURE 34. Adjusting the step size to smooth our image stack and achieve a more continuous deformation.

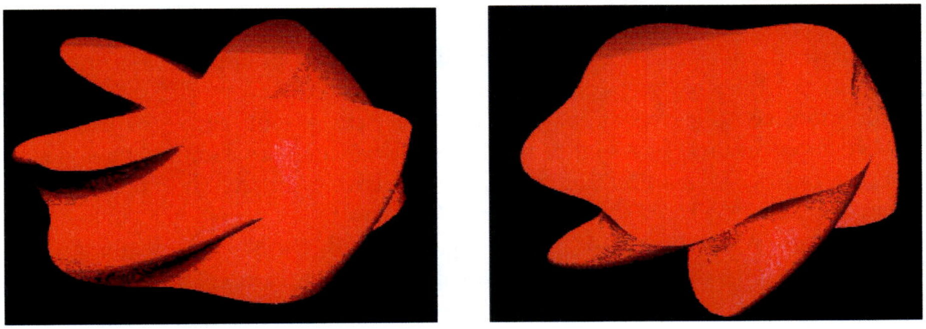

FIGURE 35. A few more angles of the smoothed deformation space.

Figure 35 shows a few other points of view of the model Chimera produced. We now export the model from Step 2 as an STL. To do this we merely export the scene, choosing the STL file extension (Figure 36). We should make sure to export the model before rotating it at all in the viewscreen. This way it will be oriented

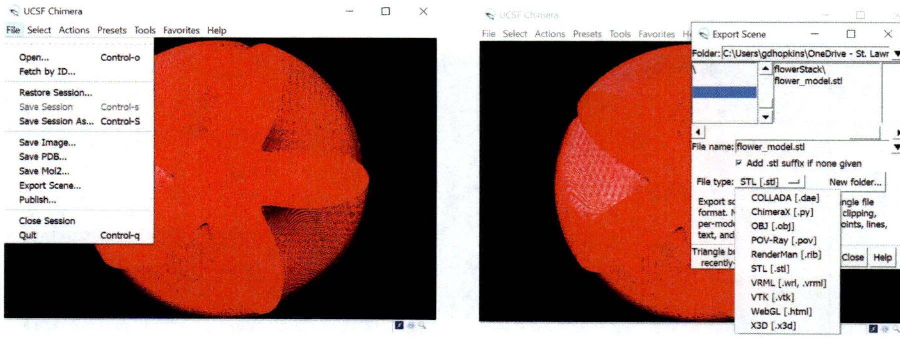

FIGURE 36. The process for exporting the deformation space.

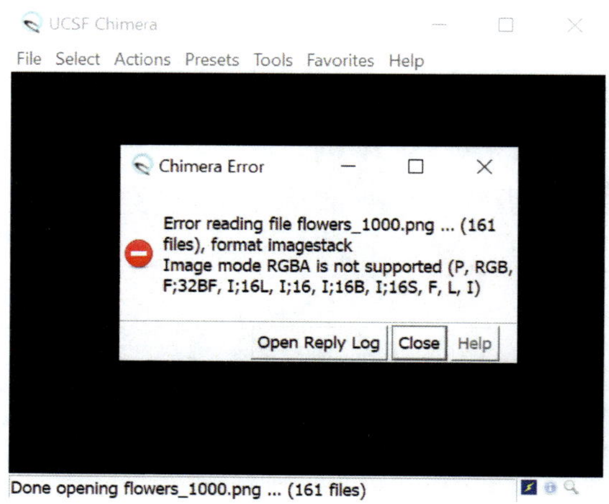

FIGURE 37. You may get this error message if you have an alpha layer. We can remove this using a python script.

correctly when opened in a slicer, lying flat against the buildplate (see the Step 3 outline above).

For more complicated animations this can take a long time—for example, if you need more frames, or a higher resolution to capture the complex behavior you are interested in. This animation, though, is relatively simple, and on the author's current machine[13] it took about 3 minutes to export.

Warning. When importing your frames to Chimera at the beginning of this step, you may have encountered an error message saying, among other things, that `RGBA is not supported` (Figure 37). This is because the frames were saved as PNG files *with an alpha layer*. To fix this, you must remove the alpha layer. One can do this in Python using the Python Imaging Library (PIL).

```
from PIL import Image
image = Image.open(fileName).convert('RGB') #conversion step
image.save(fileName)
```

[13]HP ProBook 445 G8, AMD Ryzen 5 5600U with Radeon Graphics 2.30 GHz, RAM: 16 GB

In the snippet above, `fileName` should be a path to your image. In the github repository accompanying this manuscript [**5**] we include a script, `alphaRemover.py`, which can loop through your entire image stack and remove the alpha layer from each frame. The program `flowerPlot.py` accompanying this tutorial already removes the alpha layers from your stack, so this step will only be necessary if you created your image stack elsewhere.

(Tutorial) Step 3. The model is now ready to be opened in slicing software (we use Cura [**4**]), sliced, and then sent to a printer (Figure 38).

5.3. Where do I get... Our intent with this tutorial is to have the reader follow along on their own machine. This way they learn the workflow and can integrate it into their own mathematical practice. In order to follow along, the reader will need the following tools.

- **Chimera:** At the time of writing of this document, UCSF Chimera [**25**] is available for free download here: `https://www.cgl.ucsf.edu/chimera/`.
- **Cura:** At the time of writing of this document, Untimaker Cura [**4**] is available for free download here: `https://ultimaker.com/software/ultimaker-cura`.
- **Tutorial Files:** All the files generated while making this tutorial are available at our public github repository [**5**]. This includes:
  - A python program generating the image stack.
  - A Mathematica notebook generating the image stack.
  - A python program for removing the alpha layer from a PNG stack.
  - The PNG images that serve as the frames of the animation.
  - The STL exported from Chimera.

If the reader would only like to follow along for one step, for example, stitching together the image stack in Chimera, they are welcome to download the frames from [**5**] and experiment in Chimera themselves.

6. Future directions

This article only begins to scratch the surface of what the author believes to be a vast array of possibilities for 3D printed static animations in mathematical illustration and research. Indeed, with how ubiquitous the notion of deformations are in mathematics, and how useful animations on a screen have proved to be, we look forward to exciting new directions for this technique of illustration and experimentation. We hope that with the techniques developed in Section 5, the reader will have the tools necessary both to incorporate 3D printing into their practice, and to find new applications for static animations. We'd like to conclude with a couple of potential next steps.

6.1. Exploring the Mandelbrot set. As we saw in Section 4.4, the field of holomorphic dynamics offers a rich menagerie of static animations to investigate, including many stemming from paths within the Mandelbrot set. The process for creating static animations from paths in the Mandelbrot set is already established (and can be accessed in our public github repository [**5**]), so all one needs are parametrized paths and to begin to explore.

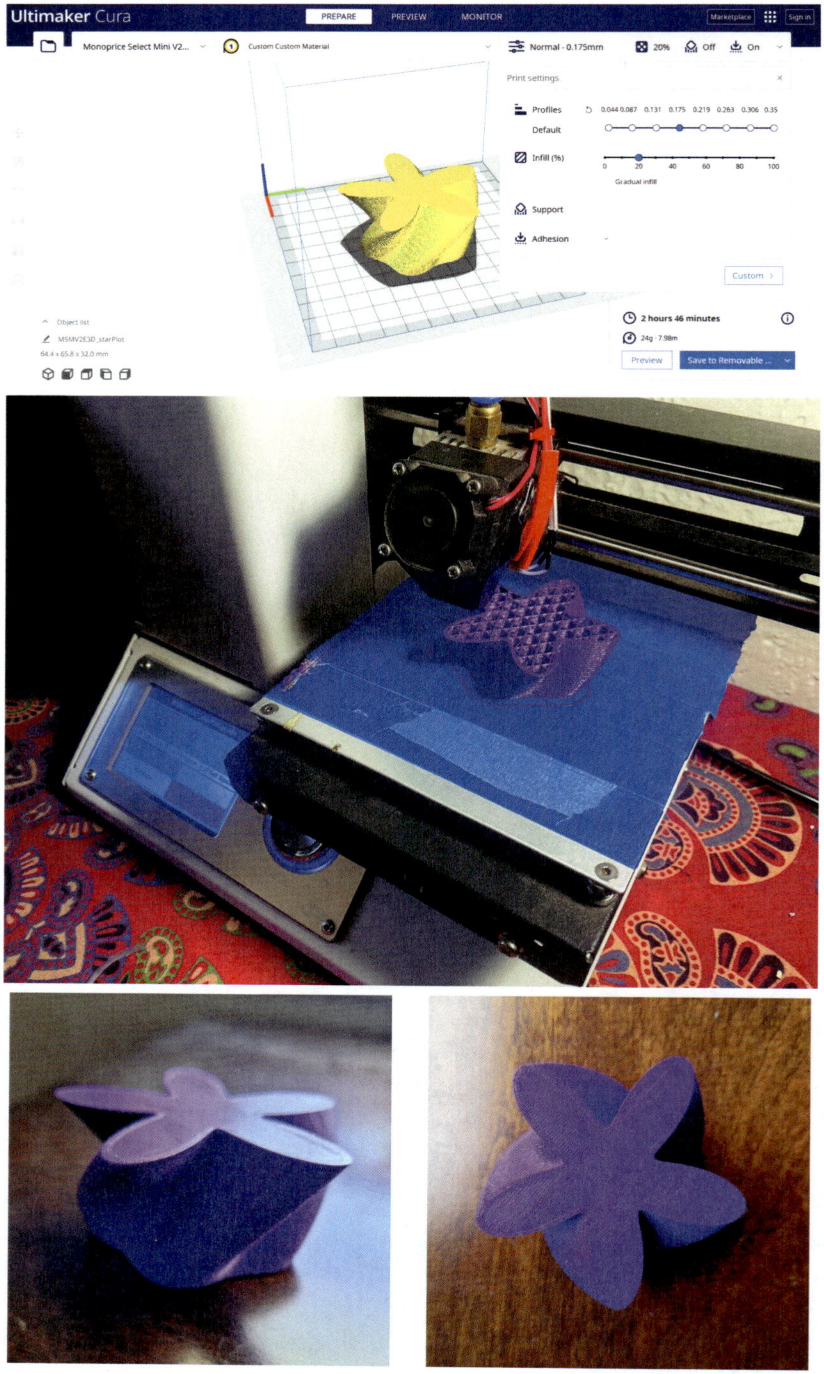

FIGURE 38. Going from the prepared file to a 3D printed solid.

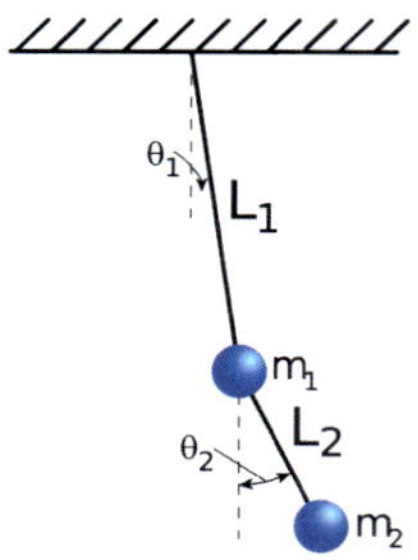

[**15**]

FIGURE 39. The double pendulum experiment.

6.2. Chaos and the double pendulum. The double pendulum [**23**] is an example of a chaotic system. In this system, attached to the mass at the end of a pendulum there is a second pendulum with another mass (Figure 39). This is a dynamical system governed by a set of ordinary differential equations which is *chaotic*, that is, it is highly sensitive to its initial conditions. Animations and experiments with double pendulums show that, even with two very similar starting points, the behavior of the moving pendulums will quickly diverge quite radically. We imagine turning the computer animation into a static animation, building the subsequent positions of the double pendulum on top of the last. We have begun such an exploration with the help of Daniel Piker using his physics engine, Kangaroo [**22**].

6.3. And more... We invite the reader to include static animations in their own investigations. Any place that a 2D animation has been useful, explore adapting it to a 3D static animation. We hope that, with the gentle learning curve and connections to many mathematical disciplines, folks will find this to be an accessible avenue to the world of 3D printing in mathematics. Indeed, as we showcased in Section 4.4.3, we have begun to see researchers lead undergraduate research investigations which use these techniques [**1**].

There are many fascinating applications of 3D printing to exploring and illustrating mathematics, and we hope that static animations will both become a part of that canon, and also serve as an invitation to the broader world of 3D printing in math.

Appendix: Deformations in algebraic geometry

The author's initial motivation to create 3D printed static animations of continuously deforming mathematical objects came from encountering these types of constructions in algebraic geometry, and more specifically within deformation theory. Here, one often will think of parameters of variation as a spacial variables and produce a deformation space of higher dimension to which witnesses the interpolation. Indeed, if C and C' are varieties which can deform to one another, exhibiting them both as living within some larger deformation space D gives a geometric framework wherein information can be passed between them via the geometry of D. As this technique was the authors inspiration to begin creating static animations, we'd like to use this appendix to describe this connection in some more detail, using examples of flat families of algebraic curves.

Suppose you are studying a family of curves C_t parametrized by $t \in k$ where k is a field (little is lost imagining k to be $\mathbb{C}$ or $\mathbb{R}$). Suppose further that each curve is the zero set in k^2 of a polynomial equation in 2 variables x and y. The key observation is that this t can be introduced as a new *spacial* variable, defining an equation, now in x, y, and t, whose zero set in k^3 we will call D. Then D comes equipped with a projection $\pi : D \to k$, which plucks out the t-coordinate, and each member of the original family of curves can be recovered as the fibers of this projection:

$$C_t = \pi^{-1}(t) \subseteq D.$$

The remarkable thing about this viewpoint is that relationships between the various curves C_t in the family can be understood in terms of the global geometry of the deformation space D, which can be studied as an algebraic variety in its own right. For example, one can use the geometry of D to show that the genus of the curves in such a family is constant [**13**, Corollary III.9.13].[14] To connect with our 3D printing philosophy, the 3D printed static animation plays the role of the total deformation space D, which can be viewed as a vertical stack of the C_t at height t. Much like how the global geometry of D provides new insights into the connections between the parametrized curves C_t, this paper begins to ask how the global structure of a 3D printed static animation reflects the connections between the parametrized objects it contains.

We give one explicit example of the deformation space of a family of curves. To do this let $k = \mathbb{R}$ and begin by considering the hyperbola $xy = 1$, and perturbing the equation, considering $xy = \frac{1}{2}$, or $xy = \frac{1}{4}$ (Figure 40). We can continue decreasing

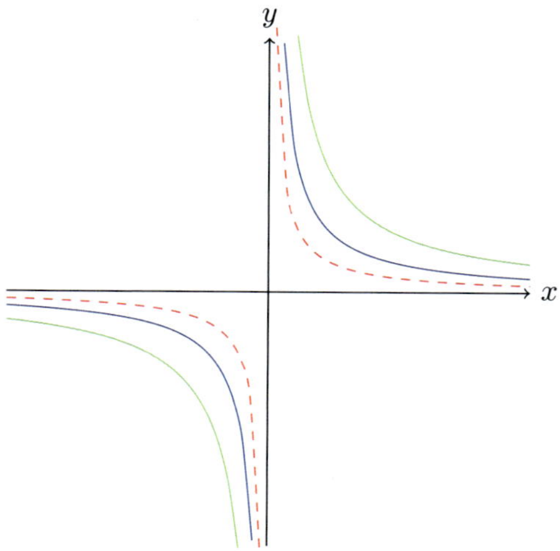

FIGURE 40. Perturbing k in $xy = t$, from $t = 1$ to $t = \frac{1}{2}$ to $t = \frac{1}{4}$.

this constant term, considering the curve C_t given by the equation $xy = t$ for $t = \frac{1}{8}, \frac{1}{16}, \dots$ (Figure 41). As t approaches zero the curves appear to be converging

[14]The genus of a curve can be defined over any field, but if $k = \mathbb{C}$, then the curve lives as a subspace of 1 complex dimenion in $\mathbb{C}^2$ and can therefore be thought of as a Riemann surface, allowing one to determine genus in the usual way

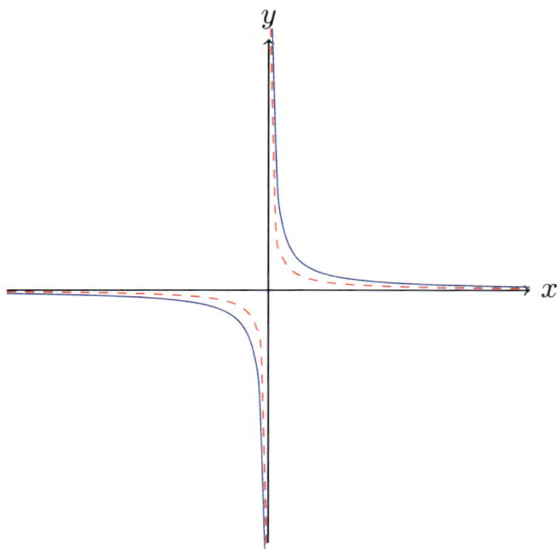

FIGURE 41. Perturbing k in $xy = t$ for smaller and smaller values of t.

to the axes. And indeed, $xy = 0$ if and only if $x = 0$ or $y = 0$, so that C_0 is precisely the union of the x and y axes. We have therefore constructed a family of curves C_t, parametrized by $t \in \mathbb{R}$, which continuously interpolate between hyperbolas and a union of 2 lines. Of course, nothing is stopping us from treating t as a *spacial* variable. That is, we can consider the the surface in $\mathbb{R}^3$ defined by the equation $xy = z$ (Figure 42). The horizontal cross section at height $z = t$ is precisely the

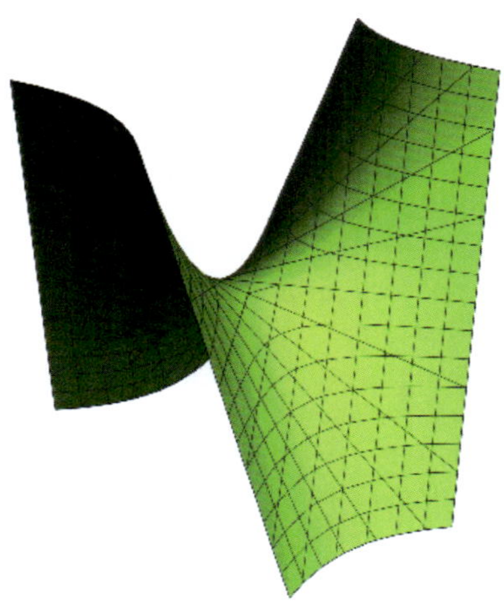

FIGURE 42. Graphing $xy = t$ in 3-space.

curve C_t given by the equation $xy = t$, and if one considers the associated contour lines, one recovers the picture of the deformation of hyperbolas to the planar axes (Figure 43). In this way, the surface $xy = z$ is a higher dimensional space which witnesses the continuous deformation of hyperbolas to a union of 2 lines.

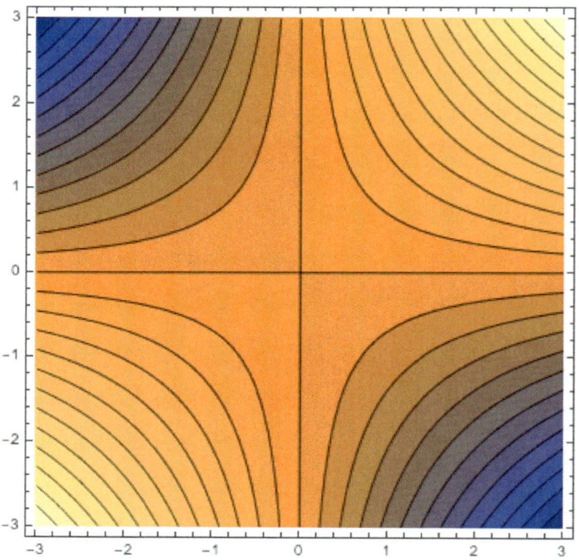

FIGURE 43. A contour diagram for $xy = z$ recovers our hyperbolas.

Acknowledgments

The author's initial exploration into static animations was as the fabrication manager of the Washington Experimental Math Lab at the University of Washington—supported by NSF grant DMS-1559860—and the author thanks Jayadev Athreya for the technology and support in exploring math from an experimental and illustrative perspective during this time. The author thanks Bernat Espigulé for his generosity and collaboration in creating static animations of Julia sets, Caroline Davis for extremely enlightening conversations, and Edmund Harriss for taking many of the photographs of their models which are included in this manuscript. We also thank the anonymous referee for many helpful comments and contributions, including some of the python code in our tutorial section, as well as the description of our process as a *reverse tomography*. Much of the development of the ideas and models were made during the 2019 ICERM semester titled *Illustrating Mathematics*—supported by NSF grant DMS-1439786—and the author thanks all the participants for their collaboration and support during that semester. The author also thanks the participants of the 2021 Illustrating Mathematics Summer School at PCMI, and the 2022 Joint Mathematics Meetings mini-course *3D Printing: Challenges and Applications*. Many of the 3D models were made with UCSF Chimera, developed by the Resource for Biocomputing, Visualization, and Informatics at the University of California, San Francisco, with support from NIH P41-GM103311. The author is currently supported by NSF grant DMS-1646385 as part of the Research Training Group in Arithmetic Geometry at UC Berkeley.

References

[1] Jasper Baltz, Caroline Davis, Bethany Mussman, and Mark Ronnenberg, *3d printing julia sets*, `https://sites.google.com/view/caroline-davis/teaching`.

[2] Arnaud Chéritat, *Techniques for computer generated pictures in complex dynamics*, https://www.math.univ-toulouse.fr/~cheritat/wiki-draw/index.php/Mandelbrot_set.

[3] Matt Conroy, *Animation of the Pythagorean tree fractal*, https://www.madandmoonly.com/doctormatt/webimages/animations/pythagoreanFractal2.gif.

[4] David Braam, Ultimaker, *Ultimaker Cura*, https://ultimaker.com/software/ultimaker-cura.

[5] Gabriel Dorfsman-Hopkins, *Deformation spaces and static animations: Accompanying code and files*, https://github.com/gdorfsmanhopkins/staticAnimations, A github repository.

[6] Gabriel Dorfsman-Hopkins, *Tutorial and demonstration (slides)*, http://www.gabrieldorfsmanhopkins.com/files/JMM3d.pdf, Part of the 2022 JMM Mini Course, *3D Printing: Challenges and Applications.*

[7] A. Douady and J. H. Hubbard, *Étude dynamique des polynômes complexes. Partie I* (French), Publications Mathématiques d'Orsay [Mathematical Publications of Orsay], vol. 84, Université de Paris-Sud, Département de Mathématiques, Orsay, 1984. MR762431

[8] A. Douady and J. H. Hubbard, *Étude dynamique des polynômes complexes. Partie II* (French), Publications Mathématiques d'Orsay [Mathematical Publications of Orsay], vol. 85, Université de Paris-Sud, Département de Mathématiques, Orsay, 1985. With the collaboration of P. Lavaurs, Tan Lei and P. Sentenac. MR812271

[9] Adrien Douady, *Does a Julia set depend continuously on the polynomial?*, Complex dynamical systems (Cincinnati, OH, 1994), Proc. Sympos. Appl. Math., vol. 49, Amer. Math. Soc., Providence, RI, 1994, pp. 91–138, DOI 10.1090/psapm/049/1315535. MR1315535

[10] Michael Dwamena, *How to use cura for beginners – step by step guide & more*, https://3dprinterly.com/how-to-use-cura-for-beginners-step-by-step-guide-more/, [Online; accessed November 12, 2022].

[11] Dante Giarrusso and Yuval Fisher, *A parameterization of the period 3 hyperbolic components of the Mandelbrot set*, Proc. Amer. Math. Soc. **123** (1995), no. 12, 3731–3737, DOI 10.2307/2161901. MR1301497

[12] Walter Hannah, *Internal rays of the Mandelbrot set*, 2006, Cornell University, Undergraduate Honors Thesis.

[13] Robin Hartshorne, *Algebraic geometry*, Graduate Texts in Mathematics, No. 52, Springer-Verlag, New York-Heidelberg, 1977. MR0463157

[14] Allen Hatcher, *Algebraic topology*, Cambridge University Press, Cambridge, 2002. MR1867354

[15] JabberWok, *Double pendulum svg*, https://en.wikipedia.org/wiki/Double_pendulum#/media/File:Double-Pendulum.svg, [Online; accessed June 28, 2022].

[16] Marius Kintel, Claire Wolf, *Openscad*, https://openscad.org/index.html.

[17] John Milnor, *Dynamics in one complex variable*, 3rd ed., Annals of Mathematics Studies, vol. 160, Princeton University Press, Princeton, NJ, 2006. MR2193309

[18] John Milnor, *Hyperbolic components*, Conformal dynamics and hyperbolic geometry, Contemp. Math., vol. 573, Amer. Math. Soc., Providence, RI, 2012, pp. 183–232, DOI 10.1090/conm/573/11428. With an appendix by A. Poirier. MR2964079

[19] 3D Now, *Cura 3d slicer for beginners*, https://www.youtube.com/watch?v=eUNTlb5pEWA, [Online; accessed November 12, 2022].

[20] Han Peters and Liz Vivas, *Parabolic implosion*, Notices Amer. Math. Soc. **67** (2020), no. 8, 1095–1103, DOI 10.1090/noti. MR4187858

[21] E. F. Pettersen, T. D. Goddard, C. C. Huang, G. S. Couch, D. M. Greenblatt, E. C. Meng, and Ferrin T. E., *UCSF Chimera–a visualization system for exploratory research and analysis*, J Comput Chem **25** (2004), no. 13, 1605–12.

[22] Daniel Piker, *Kangaroo3d*, http://kangaroo3d.com.

[23] P. H. Richter and H.-J. Scholz, *Chaos in classical mechanics: the double pendulum*, Stochastic phenomena and chaotic behaviour in complex systems (Flattnitz, 1983), Springer Ser. Synergetics, vol. 21, Springer, Berlin, 1984, pp. 86–97, DOI 10.1007/978-3-642-69591-9_9. MR771695

[24] Steven H. Strogatz, *Nonlinear dynamics and chaos*, Addison Wesley, Reading, MA, 1994.

[25] UCSF Resource for Biocomputing, Visualization, and Informatics, *Ucsf chimera*, https://www.rbvi.ucsf.edu/chimera/.

[26] Wolfram Research Incorporated, *Mathematica, Version 13.0.0*, https://www.wolfram.com/mathematica, Champaign, IL, 2021.

212 Valentine Hall, St. Lawrence University, 23 Romoda Drive, Canton, New York, 13617

Email address: gdhopkins@stlawu.edu

Proceedings of Symposia in Applied Mathematics
Volume **79**, 2023
https://doi.org/10.1090/psapm/079/00713

Making and breaking rules with clay and code: iteration, glitch, and mathematical thinking

Timea Tihanyi

Abstract. The present paper introduces 3D printing from the perspective of a visual artist's practice. It considers mathematical ideas from basic geometry and trigonometry to computational theory for the creation of forms and textures, which otherwise would not be possible to create. 3D printing with clay is a recent development within a repertoire of available additive technologies. In ceramic craft, the process of building up a form coil by coil is one of the most ancient techniques. Clay printers revisit this process with unprecedented granularity, accessibility, and transparency. The digital-to-physical workflow creates limitless entry points for using mathematics in creating a design from scratch or hacking an existing design. I introduce several of my recent projects and collaborations with mathematicians and design researchers at Slip Rabbit Studio using the Rhinoceros/Grasshopper software for form giving, and using data physicalization, coding and machine manipulations.

Creating a dialogue around cross-disciplinary collaboration

Art and math are often posed as opposites, manufacturing a false binary. New making technologies, such as 3D printing, connect the two areas of inquiry in visible ways. For both artists and mathematicians, the most exciting part of research begins in situations where existing tools no longer suffice. New methodologies, conceptualizations, terminologies, software and workflows are necessary, and they need to be invented for the specific problem. The creative work is a self-galvanizing process where the discipline boundaries often blur, and this intersection is where printing with clay currently resides.

I am a visual artist working with clay, math and technology. I've been a ceramicist for over 25 years. In 2016, I founded Slip Rabbit, a research and mentoring studio focused on digital ceramics, which opened its doors to interns, research collaborators and artist residents in the following spring. I also teach in the Interdisciplinary Visual Arts program at the University of Washington, including a co-taught course Making Meaning: Art and Math as Embodied Practice. In my artistic research, I generate experimental processes for 3D printed forms in clay.

Using computational tools and digital technologies in the ceramic process offers great opportunities to contemplate the value of the unexpected within systems that are built to be infallibly predictable. Iterative practices, play and error, and the role of failure are recurring themes in my work.

Art making is a way of thinking: through the material, tools and hands, and by rigorously analyzing and reflecting on the processes itself. In the practice of Craft, using any type of medium comes with a responsibility toward its own context and history. In other words, clay is not a prototyping material; it's a medium embedded in its own ceramic traditions and discourse. Digital technology, such as ceramic 3D printing, fits into this discourse, and I claim that it is necessary to interpret any object made with clay within this framework.

Technology has not and will not change ceramics. But technology itself changes extremely rapidly. Computation-based tools, software and hardware, including user interfaces, platforms, and workflows used today may be completely obsolete by tomorrow. Anyone beginning to work with a clay printer will find it an easy learning curve, which lends itself to innovation by experimentation. The mathematician's understanding of the underlying principles of geometry, computation, and coding will open new experimental avenues when creating 3D printed forms in clay.

I have noticed over the years that as my projects were getting more complex, the computational expenses of the digital workflow in 3D practice also increased, often exceeding my technological resources. This is one of the reasons for which I find myself in constant need of innovation and investment; theorizing, building and testing out new workflows, strategies and software. There are many ways of reaching a similar-looking outcome, but, often, there are clear aesthetic and conceptual differences as the consequence of specific design solutions. These subtle differences matter to the maker, and they will matter to the audiences too, since each digital object is a token, holding a larger narrative of its origin, purpose and place in the world. Computational design, additive manufacturing, smart objects, and techno-ceramics has expanded into vast and intersecting fields of research. Progress would not be possible without cross-disciplinary collaborations. Fostering communities where interactions between physical and digital makers can happen informs our shared learning with technology and our expectations toward it.

3D printing is an additive fabrication process. Working with design geometries in CAD (Computer Aided Design) is not so different from building with LEGO blocks. I found that the various types of 3D modeling software widely available today (with many of them being free) can be a great teaching tool for practicing complex concepts in geometry, even at the college level. Through ceramic 3D printing, the potential intersections between math, programming, engineering, architecture, design and ceramics seem inexhaustible. Since ceramic extrusion printers tend to lead to less predictable results than other common types of filament printers, printing with clay also gives teachers a playful pedagogic opportunity to foster learning by doing and to emphasize valuable connections between rigorous scientific method, artistic instinct and trial-and-error.

The digitally aided workflow creates limitless entry points for creating and hacking a design, be those on the level of the design software, code, machine properties and settings or focused on the ceramic materials themselves. In 3D printing, one can arrive to the final form through a variety of parallel paths. Because the process takes many pieces of software and equipment, many of which are still beta tested, figuring out how these multiple platforms would fit seamlessly together, has often challenged me. On the other hand, as printing is now mainstream, the design-to-object pathway is also becoming better connected with a more user-friendly and intuitive interface. During the making of the projects described here, math was sometimes a starting place, other times, it was a tool. I did not set out to illustrate a math concept, rather, my goal was to explore the inner logic of mathematical systems and how they relate to physical processes in the tangible world.

Ceramic paste extrusion printing

Ceramic paste extrusion printing is a rapidly developing area that has recently found its way into both art and design practice. Experiments with the process began around 2009 through research initiatives concerning both hardware and software possibilities. Belgian designers Dries Verbruggen and Claire Varnier of Studio Unfold [**SU22**] and the British ceramic artist Jonathan Keep [**JK22**] were among the first to explore printing with soft (plastic) paste ceramic materials and to publicize their research methodology. Early experiments around the plastic printable materials, such as concrete; resinous and organic compounds; as well as clay, were widely shared among the global community of makers. The introduction of these plastic materials within the automated digital process helped to refocus the dialogue from algorithmic perfection onto the role of imperfection.

The ceramic 3D printer I will describe below is a type of material extrusion printer, professionally referred to as Fused Filament Fabrication (FFF) or Fused Deposition Modeling (FDM) printer. It is an extruder that works with a soft paste material, which is contained in a pressurized chamber above the nozzle. Similarly to the process of knitting and weaving, FDM printing builds up the form one thread-like row at a time. Each row and each texture element affect the adjacent ones and, consequently, affects the entirety of the form. Thus, the ceramic print process combines both the structural possibilities (and challenges) of traditional clay techniques, like extrusion and coil-building. What is novel about clay printing is its granularity: building up a form by following a list of spatial coordinates amended with a sparse dictionary of machine instructions (GCODE). This 2D-to-code-to-3D translation gives the process an unprecedented transparency, including opportunities to hack every step in the making process. These hacks can happen at the level of form giving in Computer Aided Design (CAD), when exporting forms as OBJ/STL files or when slicing them into GCODE files and during machine operations.

Digital ceramics is grounded both in the world of math and in the realm of physical matter. In ceramic 3D printing, the material is dominant, working with the unexpected, leaving room for improvisation and play is a necessity. Objects constructed with algorithmic/parametric processes offer limitless iterations by simply rescaling and remapping the form or by combining volumes (Boolean operations). Every time a file is printed, it comes out slightly differently, especially if slicing or machine settings (pressure, speed, nozzle diameter) are being changed. Clay is an organic material, very much alive. With properties that change with the weather, room temperature, sourcing of the ingredients and many other factors during its manipulation, the resulting in outcome may be anticipated with practice but never guaranteed.

How do ceramic extrusion printers work?

At my studio, Slip Rabbit, we use two different ceramic FDM printers: a cartesian-type Potterbot and a delta-type WASP. Regardless of the developer or the type, any extrusion-based printer existing on the market today consists of a tank, which holds a certain volume of clay under pressure; a build-platform; stepper motors, each of which allows movement on one of the three axes; one additional stepper that is responsible for the extrusion–directly or through an auger system– and a nozzle, which reduces the flow to the desired diameter. The biggest difference among the functionality of these printers (not counting the obvious size and volume limitations), is the readiness to retract, meaning that when the nozzle disengages, the flow of clay immediately stops. Since each printer model is slightly different, the most exciting consequence of the previous statement is that each machine has its own unique "fingerprint" on the finished object.

Limitations of the tool

3D printing is often promoted as an automated process: push the start button and walk away. While this is rarely true in general, in ceramic extrusion printing one must consider the material. Rather than being an omnipotent maker-machine that automatically materializes the object on the screen, it is more helpful to think about the extrusion printer as an expansion of a very traditional pottery tool. It is indeed a nimble and smart clay extruder that can follow location directions and respond with pressure and speed changes. While it is possible to get a near-exact object printed, the most interesting potential of ceramic printing is in the unexpected, in the mishaps. Rather than denying, it makes sense to take advantage of the unpredictable nature of this unique material. Not only letting it to do what it can do best, but facilitating forms that can buckle, slump, crack and crease, sometimes elegantly, other times rather dumbly. Clay sticks best to clay, layers won't build over void, gravity must be respected. These are the most important things to keep in mind.

Rules and systems in art and math

Artists, like the conceptualist Sol LeWitt [**LeW19**], who explicitly created work by combining simple units of lines, colors or shapes in a rational manner playing out all possible variations, have been exploring the contemporary paradigm of iteration and algorithmic thinking. But, for thousands of years before these contemporaries, generations of artists in every culture of the world have created meaningful designs by setting up simple sets of rules and iterating them. Weavers, basket makers, beaders, and embroiderers innovated with repeating units, organizing them into complex systems of motifs. The common element in each of these practices is playful improvisation based on serious mathematical thinking. Often described as algorithmic in nature, the goal of this work has been focused on cultural meaning through abstraction. The resulting designs have been analyzed and classified by mathematicians, while the making process and the tools themselves, which reveal the creativity and problem solving ability of the individual maker remain to be further explored and taught along with the patterns.

As a maker, I am invested in experimenting with processes, tools and strategies, and sharing these with the public. I do not claim any formal education in mathematics, so when describing my projects and processes below, I will use the term **math** to refer to *mathematical thinking*, while reserving the term **mathematics** to refer to the established *area of study*. My interest in math started with an interest in designing and building complex systems using only a small set of repeating units. Varying these by applying a very limited set of rules initiated a collaboration with Sara Billey. Our shared inquiry considered discrete processes with self-organized behavior such as the elementary cellular automata, chip-firing games and sandpile groups, all of which produce almost limitless combinations of patterns. These, like weaving and knit patterns, share basic structures (cells and lattices), basic behaviors (a finite number of discrete states), and rules, which are applied iteratively. Repeating these rules leads to a specific arrangement of states (for example, on/off) for each unit, all of which together form a matrix. In this matrix, regularities and relationships are recognized both locally (within a small area of neighboring cells) and globally (across the "game board"). This is what our human cognitive ability recognizes as a **pattern**.

Working with the elementary cellular automata

Exploring the elementary cellular automata gave me the first opportunity to develop and test methods for turning a numerical matrix into a printed ceramic texture. An elementary cellular automaton pattern develops as each row (from generation 0 to generation n) is transcribed into the next consequent generation by applying one of the $2^8 = 256$ possible rules. The rows are made of a single strand of cells, each cell taking on one of the possible binary states (on/off). The state of each cell in the new generation is determined by the state of the original cell and the states of its two neighboring cells in the previous generation, leading to eight possible binary states for the triplet [**Wol02**]. Very few of these rules actually leads to an interesting pattern, most of them are fairly predictable and stable throughout hundreds of generations. Mapping these rows and cells onto the structure of a CAD model took some trial and error, necessitating a reverse engineering of the process. To generate the 2D pattern plots, I used a Python code written by Billey in CoCalc, which ran the selected rule for the specified number of generations. I learned to manipulate this code, changing and altering according to my actual research questions and artistic goals. One of the most fascinating of these experiments was the writing of extraneous rules into the original code. These made-up rules, for example, one that responded to the occurrence of a certain pattern by flipping the states of the target cells in the generation below that, prompted a dramatic cascade of change all over in the resulting matrix. Working on this in 2018, we had no idea yet of the uncanny resemblance of our models to what a year later became the behavior of the virus spreading in susceptible communities. The 2D plot of the resulting matrix from CoCalc was then imported into Rhinoceros, a computer-aided design application software, developed by Robert McNeel & Associates, where it was reconstructed as a series of planar volumetric objects. Working around a surface with a constant radius was the best method that offered itself for remapping the 2D matrix plot in the 3D world, so I stated working with cylinders. Keeping the form simple, also allowed one to focus on the resulting texture, which have emerged in interesting ways, based on scaling, spacing, layer height, and nozzle diameter. We also had to decide what to do about the edges, and this is where the efficiency of the computational process was the most useful. Wrapping the 2D matrix around the cylindrical form, it made good sense to consider each row as a closed loop with no beginning or end.

I worked back and forth between the STL file and the printed model: changing model, slicing and print settings until I was able to dial in a result that was not only visual but also tactile. This type of work relies on over-extrusion, deliberately forcing the printer to leave the perimeter of the object, resulting in an unsupported clay loop or spur. Spacing these carefully and strategically is always a result of lots of calculation, and even more trial and error. It's very important to emphasize, how much this work depends on the actual experimentation and making of intentional mistakes. The correct speed, amount of pressure, the right length and density of texture are questions as much aesthetic as technical in nature.

Work with abelian sandpile models

The work with the elementary cellular automata was followed by a year-long collaborative exploration on Abelian sandpile models, which also involved a group of undergraduate and graduate students in mathematics, members of the UW eXperimental Mathematics Lab (WXML). Sandpile models are a form of cellular automata on a finite grid, where each cell has an associated value that corresponds to the pile of sand the cell can hold. Sandpile models have been studied in Combinatorics and Graph Theory as well as from perspectives outside of mathematics. In fact, this model was first proposed in physics by Per Bak, Chao Tang and Kurt Wiesenfeld [**BTW87**] as an example of a Dynamical System displaying self-organized criticality. Bak argued that small disturbances can lead to major events in complex systems through the evolution of a critical state, such as sand piled high and ready to topple. In the critical state, the interaction of the whole matters, rather than individual and localized actions of each grain and each cell that make it up. While many open problems in this area of mathematical research are still unsolved, the fundamental concept of the sandpiles coming to tipping point and toppling sand from one cell into its neighboring cells is quite accessible visual model for general audiences. The rules of the game are fewer, but the possible matrix plots are much more unpredictable. This was fascinating to me not only for the beautiful mandala-like visual patterns but also for being a nimbler and more sensitive math strategy for modeling Nature.

Our WXML research group on sandpile models experimented with several possible sand-toppling scenarios using both regular square grids and hexagonal grids. Working through the production issues of reconceptualizing the 2D sandpile model plots as 3D forms, also brought up many interesting questions about the mathematical behavior of these models. For printing, the size of the game board had to be determined by the number of layers in the finished object and the circumference of the finished object in relationship to the printing nozzle. In clay extrusion printing, layer height is usually about 1-2mm, while plastic FDM printers typically are set for 0.1-0.2mm layer heights. The work envelope of the printer allowed a circumference of 600-700mm, which using a 2mm nozzle provided about 300-350 distinct points for extrusion. While our team plotted matrices of hundreds of thousand of cells and poured in millions of sand grains, going beyond a 300x300 grid in the clay printing process was neither computationally practical, nor physically attainable. At the computational level, we exhausted capacity with larger matrices, leading to a frustratingly extended waiting period before the plot appeared. To cut down on the computing time, we identified various symmetries within the pattern and calculated only the individual section, seamlessly stitching the slices back together in the finished plot. While mostly working with a finite lattice, bounded by sinks (where the sand is able to leave the board) on every side, we also experimented with a scenario that placed the sink onto a single cell in the center. Joining the right edge of the board to the left, and the top to the bottom, this effectively turning the board inside out, leading to fascinating pattern variations.

One of the 3D printing technical problems I encountered in this project had to do with the number of possible states in a sandpile model. While the binary system of the elementary cellular automata offered an easy solution of texture/no texture (extrusion/no extrusion), each cell in the sandpile plot can take multiple discrete values. These can be easily represented in 2D by a multi-color map but require some flexible interpretation in 3D. Rendering a specific extrusion length to each discrete value was a good working solution to the problem, and I was able to vary which extrusion length was associated with a certain numerical value, producing multiple iterations of the pattern this way. While constructing these 3D matrices in Rhino, it occurred to me to analyze the specific distribution pattern for each value, which led us to devising an odometer. This counter analyzed the rhythm and the pattern by which piles toppled and the distribution of the sand changed in major ways. The students experimented with known integer sequences in order to predict the occurrences of critical states.

Beyond the mathematics, patterns of math thinking

The work with mathematical ideas led me to my current work with textile patterns. Algorithmic processes, symmetries and transformations have always had a great importance in the creation of textiles, though these were often applied as strategies of efficiency and improvisation. An embellished cloth is destined to function in the world as a covering of 3D volumes, which could be walked around and seen from all directions. Textile patterns represent the symmetries and hierarchies of bodies they are draped on. They were meant to be taken in by the eye at once and by linear progression, simultaneously. Every culture and every time period have produced a rich archive of pattern variations, many of which, we assume, have developed by individual artists innovating one small step at a time, both preserving a cultural tradition and, at the same time, remaking that anew. Women have often been in the forefront of this work, producing woven, knit and embroidered cloth embellished with complex shapes, textures, and color patterns. These were often achieved by creatively varying a binary system, such as under/over sequences on a loom or knit/purr sequences in knitting. I have been researching folk cross-stitch patterns from my native Hungary and reimagining them as porcelain forms and heavily textured surfaces. In this work, I start out with a 2D plot of a motif, which I usually find in old knitting or weaving pattern books and redraw in a binary form. In Rhinoceros, it is possible to work using layers, overlapping and superimposing them. Each layer is filled with units modified and tessellated by applying different rules. Moreover, by applying symmetry transformations to small segments of the original pattern or even to the entire layer, one can build complexity in the software, similarly to the process of the cellular automata and chip firing games. Turning on or off the layers reveals new unexpected pattern variations, which are finite and discrete dynamical systems. Mathematics is not the goal in this work, but math thinking remains part of my toolkit as I manipulate the designs into complex surfaces of pattern. In this kind of work, there is a lot of algorithmic determinism, much of which is part of the black box of the software. I relish the few but significant chance driven parts of the process. These may happen in the CAD program as I rotate, reflect and translate the design fragments, layering them upon one another.

As I watch the pattern evolve, I edit that freely for a more interesting visual effect. Little blips of chance and accidents also play a large part in printing and in the ceramic process following that. To pay homage to the textile origin of this work, I intentionally started deforming the perfect cylindrical object, first by hand, later by the drying and firing process. When clay goes through a crystalline transformation in the heat of the kiln, the form gets liberated from its rigid origins, arriving to a final shape determined by complex interactions of the material, form, heat, gravity and time.

Entangled in Fourier series

Being interested in the structure of woven and knit textiles, though for entirely different reasons, spurred a collaboration with Frank Farris. Farris uses Fourier series in a Rhino plug-in, Grasshopper, to model the loop structure of chain mail [**Far20**]. During our collaboration, we explored three basic regular knits: stockinette, garter and ribbing. Farris' Grasshopper definition approximated the binary topology of knit and purl stitches in knitting as curves described by sine and cosine functions. To make the 3D printed form viable, we bent the flat knit into a circular form. We experimented with various proportions and combinations of loops by testing them on the clay printer. Being able to use a parametric design tool was essential for creating design iterations in this work. There were mathematical, structural and manufacturing issues to be resolved, and the best way to sort them out was by making and remaking. The Grasshopper definition visualizes, then "bakes" a mesh into the Rhino workspace. All looks good on the screen but, in parametric modeling, the result is frequently a non-watertight mesh. Leaky (non-manifold) meshes are not directly printable and require repair. For fixing a mesh (STL) file, I use both Meshmixer and Meshlab, though recently some slicer programs are also capable of doing this. Mesh software is designed to fill in and heal gaps by adding missing facets and stitching seams together. They are able to create a solid, though some distortion can be expected. In a knit, the most significant structural characteristic is the negative space among the loops. This gives the knit its visual beauty. Of course, this three-dimensional lattice poses quite a challenge for the ceramic medium and for the printer. For comparison, I experimented with slicing the same STL file for both a plastic FDM printer, a Prusa i3 MK3S+ using PLA, and for a WASP10040 ceramic printer. The resulting objects looked extremely different. Scale change (layer height 0.15-0.2mm for the Prusa and 1-1.5mm for the WASP, and nozzle diameter 0.4mm for Prusa and 2-3.5mm for WASP) created different proportions of mass and void in the two materials, dramatically changing the need for external supports and retraction values. Most importantly, the physical properties of the material itself created vastly distinct outcomes on the given scale, many of which would benefit from further experimentation and research.

From data to form

Even elementary math concepts can bring interesting results when used as hacks in the process of 3D printing. In the summer of 2018, I collaborated with IxDesign researcher, Audrey Desjardins on a series of drinking vessels, called ListeningCups. In the making of these, we used ambient sound, which we had gathered from a variety of locations. For the ListeningCups, I devised a way for the 3D printer to transcribe sound data into textures, not by moving in a certain predetermined way but by coming to a stop and dwelling in place for a specified duration of time. We divided our sound data into decibel ranges based on the human hearing ability and assigned a length of time to each decibel range for which the printer would pause. At chosen coordinate points in the GCODE, these pauses were added, utilizing Excel tables to combine the two sets of code. The type of clay printer I used for this project relies on continuous extrusion and has no retraction. As a result, dwelling in place creates an accidental extrusion of clay at the specified coordinate point, showing up as a noticeable bump both inside and outside of the wall. After some experimentation, we successfully correlated sound volume changes with distinguishable textures ranging from fine to pronounced. The resulting objects encapsulate audible information into tactile material. See a detailed description of this project and our definitions for the terms data tactility and data stories, in [**DT19**].

Inspired by these results, I continued developing methods based on hacking the machine and the GCODE, creating new textures, and putting math and technology in service of the material. One of these GCODE hacking projects used principles of trigonometry. Working with a Python code, which took two inputs, a GCODE of a simple form and a sound file, I was able to recalculate sine and cosine of the original spatial coordinates and move them to a specified distance in one of four randomly assigned new directions. At regular intervals in the GCODE, sound information would send the printhead to a revised location, creating a unique alternative path each time from one coordinate point to the next. The resulting extrusions weave in and out of the surface of the vessel, each sound environment making its unpredictable footprint through rather rational means in clay. The result is an organic form, much like a bird's nest, built by math and chance.

Setting up a clay print

As part of the documentation of a commercially available 3D printer, a "printer profile" is usually made available by the manufacturer. This profile is imported into the slicer software and will act as a destination-specific translator when creating the GCODE. I tend to make modifications to this basic profile quite often, especially for the WASP printer. Adjustments respond to each specific printing situation: in addition to differences originating from the nozzle diameter, there may be significant changes necessitated by clay consistency or the complexity of the design.

Each new print object requires the re-checking and changing of the actual print settings in detail. These include layer height, infill, top and bottom solid layers, number of shells, and additions like a skirt or a raft. For a 3.5mm nozzle on the Potterbot printer, I typically use anywhere between 1-1.2mm layer height, 4-4.5 extrusion multiplier, 1 shell, 2 or 3 layers of bottom and 1 skirt layer. These settings would print a fine cup-size object, but major scale changes will always require me to rethink and alter these variables. There is no right or wrong here. I take copious notes on each piece I make; and save the tested-and-true printer profile under a new name in the slicer's library of settings.

What base to print on with clay can be a challenge. Over the years, I found the 1/4" thick plasterboard as the most reliable base for the print. Various removable materials from a damp sheet of craft paper to sheets of high-density EVA foam and various other plastics would also work. When beginning a print, the printer's Z height must be calibrated in relationship to the height of the board on the platform. For clay, slowing down the printing speed to 60% in the first layer should provide good adhesion. I always set up the print with at least one skirt layer, in order to prime the nozzle and ensure sufficient flow. When a print is too difficult to start, I sometimes turn to making a flat base from a clay slab. This holds the base coil very securely but may be more difficult to remove later. The rest of the printing process is a matter of experimentation with flow, speed, and additional supports, each of which can usually be controlled during the printing. While in plastic printing it is quite common to have presets in the slicing program determine supports both inside and around the form, working with clay necessitates supports of a different nature: ad-hoc foam blocks and lumps of soft clay. The printed form will stay soft and malleable during printing and for a length of time after that, making forms with undercuts and slopes over 15% difficult to produce without design alterations.

Finishing a printed clay object

Clay continues to shrink as it dries. Because of shrinking, it is important to remove any rigid support as soon as the walls themselves are firm enough to stand on their own. A slow drying process is important for most clay bodies; and all moisture must leave the clay by the time it is loaded into the kiln. Printed clay is made by a layering process, having generally weaker adhesion between the layers than among the particles inside of a layer. Consequently, the object is less likely to explode than to gently split during the firing, should there be any residual air or moisture left in it. Structural tension may build inside of larger forms, only to reveal itself in the firing. Each ceramic form has its particular set of issues that need a unique resolution, but all can be solved with patience and experimentation.

I fire all my pieces in a computer controlled electric kiln to temperatures above 2100 degrees Fahrenheit. Preparation and application of glazes, and the firing itself, also requires a properly ventilated space away from the printers and computers, as well as personal safety equipment and a bit of experience. While printing may be done anywhere, firing and glazing are aspects of clay printing that I recommend doing in a dedicated ceramics studio. Clay facilities with workspace rentals or for-hire finishing are available in most cities.

The importance of being surprised

A ceramic extrusion printer is neither a reliable prototyping tool in design nor an efficient production tool in pottery practice. Its greatest value resides in the experimental and creative practice. It's a tool with multiple entry points: CAD, math, code, machine, structure, form and material. Throughout these years of printing, I have kept a series of notebooks. Not the typical artist sketchbooks filled with drawings of things to make, these are logbooks of making with dutifully recorded machine settings, measurements, software commands. These records of each project also contain notes on the digital to physical workflow, marking which ideas to expand on next and asking where else I may challenge my own established procedures.

I never expect the first print to work out. Every piece is provisional, and I keep testing until I have exactly what I want or feel that I can make reasonably correct predictions. There are numerous instances when the software would produce a convincing rendering, but the digital object is not viable in the physical world. Structural issues aside, what we make and why we make it are questions that need to be both posed and answered, and this is especially true for digitally facilitated objects. How do we create a place for these algorithmic, mathematical or data-driven forms in our world of experience? How do we make sense of a world that is filled with technological, algorithmic objects? Challenges posed by digital objects and the haptic paradigm change are also an invitation. Both artists and mathematicians are critical participants in responding to these questions of technological futures. There are structural, conceptual and contextual debates around the digital object, to which art-math collaborations may serve as a viable response.

Note on the role of failure

During the making of these projects, I was constantly running into various obstacles due to limitations of one kind of another: computing power, software, machine, material or even my skillsets. In all maker practices, we speak of craft, which usually refers to an experience-driven iterative procedure that keeps fine-tuning the process until the desired result is achieved. Repetition builds muscle memory and builds an inventory of possible outcomes, among which the desired result resides. Iteration nurtures accurate predictions, while the unexpected often initiate new paths of discovery. In this sense of learning by doing, digital making fits into the tradition of craft practice.

Building a large volume from delicate coils takes a risk of physics. Air bubbles may disrupt the flow, wet forms slump just enough for the next layer to miss alignment, soft clay may buckle under its own weight. These are rarely pleasant surprises. Predictability is equally dependent on craft and on sheer luck.

So why print with clay then?

Clay is the most direct, unmediated material, one that is able to build anything from nanostructures to buildings and coral reefs. Hands-on interaction is the key in getting to know clay. Potters refer to the "memory" and "personality" of clay, the material's structural intelligence (or shall we say stubbornness?) by which it remembers touch and tool.

Exactly these–clay's clumsiness, versatility, responsiveness and memory–that make ceramic printing so exciting. And exactly this unstable ground is where the opportunity to make mistakes, to fail, splendidly and intelligently, arises. Failure is a frame of mind. I trust failure to be the precursor of fresher, better ideas and more elegant solutions.

Breaking the rules with clay

Creating a clay object with a rigorous algorithmic design requires careful planning and tweaking. Assuming that all the calculations were right, we have a design file, then a machine code, finally ready for the printer. Many aspects of how math is encountered in physical form is ultimately decided by the numerous choices a craftsperson makes throughout the workflow, as well as by their individual vision for the desirable outcome. Individually, these choices are often too subtle to have an immediate visible effect. Rather, the sum of these choices will create a unique combination, which ultimately defines the shape of the resulting object.

When I started 3D printing with porcelain in 2017, the process was new and consensus predicted that this material—generally too fussy and difficult to work with—was not suitable for that process. I was warned that I would not get too far with it, which made me want to use it even more. My experiments, which started with building textures from sequenced loops, proved otherwise. I continued on more ambitious on scale and complexity, changing every variable I could think of: slicing thicker and thinner, increasing extrusion speed, switching nozzle diameters, putting the texture details closer together and farther apart. Gradually, my designs became more intricate, playing now freely with the way loops were stacked, intersected or combined to form larger clusters. Through these experiments, material and structural challenges arose, but were always successfully solved. Soon enough, I was able to create not only structurally sound walls, but also to make them very thin and translucent.

By now, most universities and high schools have ceramic 3D printers within their makerspaces and ceramic programs. The tool has become affordable and easier to use. A considerable library of visual examples and making protocols is available freely online. The dialogue, within which clay printing exists now, is vibrant and exciting. Several Hand-Machine Research Labs exist in the United States, where experimentation with clay printers in the expanded field is a stated goal. Designers send files around the world to various printers to be produced with local clays. Mathematics is a fundamental part of this digitally enabled algorithmically grounded process of form giving. The field is awaiting artist-mathematicians to join the fun.

Experimentation is a necessity. What will or will not work may be anticipated but only repeated test prints will give me a confidence of understanding why and how it works. Clay challenges all logical and predictable outcomes. Its own materiality is made even richer by a long ceramics history that is connected to our humanity, the mundane and the abject. The ceramic process leads to tactile and tangible objects with a spatial presence, relating our abstract models of complex systems back to the simple bodily experience of touch. Each finished piece has many stories, meaning accumulating layer upon layer. Mathematical ideas and craft are part origin story of 3D printed objects in clay. As an artist, it is my job is to weave a narrative that explores not only **what** these objects are, but also **how** and **why** they have come to be. Ultimately, it will always be up to the viewer's interpretation to give life, relevance and context to 3D printed objects in clay and, through them, to math ideas.

Acknowledgments

My work does not happen in isolation. The mathematical research about sandpile models has grown out of hard work by a Washington eXperimental Mathematics Lab (WXML) research group of math undergraduate and graduate students, Connor Ahlbach, Catherine Babecki, Eli Johnson and James Pedersen, lead by Sara Billey. Slip Rabbit interns, Daria Mikovic, Erika Lee, Annabelle Wu, Alison Gray, Wanna Huang, among others, played an important part in the development of many early projects. Frank Farris and I built a fruitful remote collaboration by sending files and objects to one another following a 2019 special semester in Illustrating Mathematics at ICERM [**ICERM19**], a unique opportunity, which was organized by several mathematicians and made possible by ICERM's mission to support and "expand the use of computational and experimental methods in mathematics." Furthermore, my colleague at the University of Washington, Seattle, IxDesign researcher Audrey Desjardins, and I have been collaborating since 2018 on data-physicalization and data stories.

I would like to thank Maria Trnkova and Andrew Yarmola for inviting me to speak at the AMS Short Course 3D Printing and Applications, Sándor Kovács for help with LaTeX formatting, and the referee for a detailed and generous report.

Timea Tihanyi is an interdisciplinary visual artist and ceramicist with an active research and exhibition practice spanning over three decades. Tihanyi holds an MD from Semmelweis University, Budapest, Hungary, specializing in neuropsychology. She received a BFA in ceramics from Massachusetts College of Art, Boston and an MFA in ceramics from University of Washington, Seattle, where she is a Teaching Professor in Interdisciplinary Visual Arts. Tihanyi's work has been exhibited in the United States, Brazil, Australia, Denmark, Spain and the Netherlands, including Shepparton Art Museum, Henry Art Gallery, Bellevue Arts Museum, Mint Museum of Art and Design, Society for Contemporary Craft in Pittsburg, Clay Center for the Arts and Sciences, Foundry Art Center, Museum of Glass, and International Museum of Surgical Science. Tihanyi is the recipient of the 2018 Neddy Award in Open Media, as well as many other recognitions and awards, including a Mellon Foundation Fellowship for developing a course in art and mathematics. Tihanyi is the founder and director of the Seattle-based technoceramics research studio, **Slip Rabbit**, and frequently collaborates with design researchers and mathematicians.

FIGURE 1. Sunday's Tablecloth. Cloth-like 3D printed and altered porcelain based on embroidery patterns. Ceramic pigments. 2020. 40"x20"x20" Photo: Timea Tihanyi

FIGURE 2. Potterbot clay 3D printer in action. Slip Rabbit Studio. Photo: Timea Tihanyi

FIGURE 3. Burst and Follow series. Elementary Cellular Automata patterns using various rules. 3D printed porcelain. 2018. 11"x8"x8". Photo: Timea Tihanyi

FIGURE 4. Control and Release. Installation at Linda Hodges Gallery, Seattle. 2018 3D printed porcelain and vinyl using rule 105 of the Elementary Cellular Automaton. Photo: Timea Tihanyi

FIGURE 5. Close up of the texture of Elementary Cellular Automata patterns. Printed with a 3.5mm nozzle in porcelain on a Potterbot clay printer. Photo: Timea Tihanyi

FIGURE 6. Burst and Follow Viral version. 2018. 28"x28"x10" Seeding extraneous rules in the Elementary Cellular Automaton pattern changed the predictability of that. Photo: Timea Tihanyi

FIGURE 7. Sandpile distribution patterns separated by value. Collaboration with Sara Billey and the WXML team. 3D printed porcelain. 2019. 11"x8"x8" Photo: Timea Tihanyi

FIGURE 8. Sandpile pattern. Collaboration with Sara Billey and the WXML team. 3D printed porcelain. 2019. 4.5"x8"x8". Photo: Timea Tihanyi

FIGURE 9. 3D printed porcelain using digitally manipulated Hungarian folk embroidery patterns. Ceramic pigments. 2020. 13"x10"x10". Photo: Timea Tihanyi

FIGURE 10. My Tulip World (Tulipán Világom). 3D printed and altered porcelain based on embroidery patterns. Ceramic pigments. 2020. 20"x14"x14" Photo: Timea Tihanyi

FIGURE 11. Object Permanence (detail). Porcelain and pigmented porcelain, custom wallpaper. 2022. 48"x24"x24" Photo: Timea Tihanyi

FIGURE 12. Knit collaboration with Frank Farris. 3D printed porcelain. 2020. 4.5"x8"x8" Photo: Timea Tihanyi

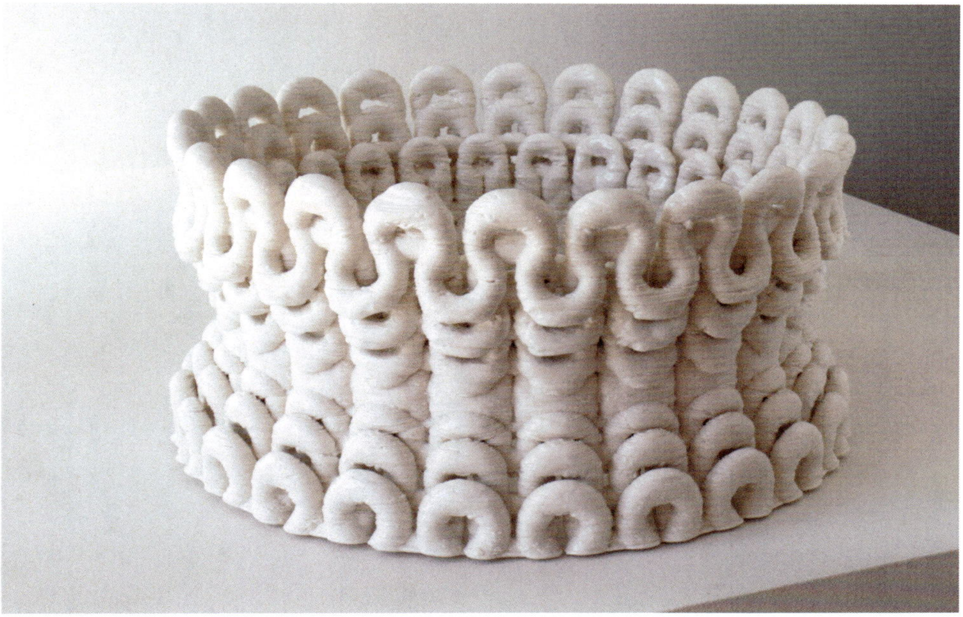

FIGURE 13. Knit collaboration with Frank Farris. 3D printed porcelain. 2020. 18"x25"x25" Photo: Timea Tihanyi

FIGURE 14. ListeningCups collaboration with Audrey Desjardins. Sound data creates texture by stalling the printer. 2018. Photo: Timea Tihanyi

FIGURE 15. Texture detail of Pathfinder. 2018. Photo: Timea Tihanyi

FIGURE 16. Pathfinder. 3D printed porcelain. Using a Python code, sound data detours the nozzle to new coordinate points creating texture. 2018. Photo: Timea Tihanyi

References

[ICERM19] *Illustrating mathematics*, Special semester program at ICERM, Sep 4–Dec 6, 2019, https://icerm.brown.edu/programs/sp-f19/.

[BTW87] Per Bak, Chao Tang, and Kurt Wiesenfeld, *Self-organized criticality*, Phys. Rev. A (3) **38** (1988), no. 1, 364–374, DOI 10.1103/PhysRevA.38.364. MR949160

[DT19] A. Desjardins and T. Tihanyi, *ListeningCups: A case of data tactility and data stories*, ACM Proceedings, Designing Interactive Systems (DIS), ACM Digital Library, 2019.

[Far20] F. Farris, *Wallpaper patterns from nonplanar chain mail links*, Proceedings of Bridges 2020: Mathematics, Art, Music, Architecture, Education, Culture (Phoenix, Arizona) (C. Yackel, R. Bosch, E. Torrence, and K. Fenyvesi, eds.), Tessellations Publishing, 2020, pp. 183–190.

[JK22] J. Keep, 2022, http://www.keep-art.co.uk/.

[LeW19] S. LeWitt, *Four basic kinds of lines & colour*, Printed Matter, Inc., 2019.

[SU22] D. Verbruggen and C. Varnier, *Studio unfold*, 2022, http://unfold.be/.

[Wol02] Stephen Wolfram, *A new kind of science*, Wolfram Media, Inc., Champaign, IL, 2002. MR1920418

School of Art+Art History+Design, University of Washington, Seattle, Washington 98195

Slip Rabbit Studio
Email address: sliprabbit@gmail.com
URL: www.sliprabbit.org, www.timeatihanyi.com

Proceedings of Symposia in Applied Mathematics
Volume **79**, 2023
https://doi.org/10.1090/psapm/079/00714

Manipulative calculus: active learning with 3D models

Janet Chen, Kelly Delp, and Stepan Paul

ABSTRACT. Manipulative Calculus is a curriculum development project focused on creating active learning lessons in calculus courses centered around digitally fabricated 3D models that we put into students' hands. In these lessons, students are asked to make geometric sense of the concepts learned in the course through problems requiring them to handle, discuss, and sketch on the models. As of this writing, the lessons and models have been in use for several semesters by dozens of instructors and thousands of students. This paper reports on the development of the 3D models and lessons for multivariable calculus, experiences and assessments of their use in the classroom, and efforts to make the curricular materials accessible to other instructors.

1. Introduction

The Manipulative Calculus project originated at Harvard University as a curriculum development project for Math 21a, a non-proof-based multivariable calculus course for a general audience. Many of the topics in the course—graphs of functions, directional derivatives, gradients, triple integrals—can be represented geometrically in three dimensions, and the project aims to help students develop geometric understanding of the concepts by designing lessons based around student interactions with 3D models. The first and third authors developed a series of lesson activities and accompanying digitally fabricated 3D models to target specific learning goals in the course. In these activities, students draw on, handle, and solve problems using the 3D models in small groups.

The activities and models were designed and produced in 2018 at Harvard University, and they have since been in use in Math 21a. Starting in 2021, the second author adopted some of the Manipulative Calculus materials for use at Cornell University in teaching Math 1920 Multivariable Calculus for Engineers. Table 1 gives an overview of the course structure of multivariable calculus at the two universities.

Our aims in this paper are to share the Manipulative Calculus materials with instructors who may be interested in adopting them and to share our experiences in developing the project with instructors who may be interested in designing or using

2020 *Mathematics Subject Classification.* Primary 97U60, 97I60; Secondary 97D40.

The authors would like to acknowledge funding for this project from the Harvard University Office of Undergraduate Education Course Improvement grant and the Cornell University Center for Teaching Innovation Micro Grants Program, as well as support from Ultimaker.

TABLE 1. Course structure for multivariable calculus courses at Harvard and Cornell and overview of how Manipulative Calculus (MC) materials were used.

	Harvard: Math 21a	**Cornell: Math 1920**
Students per semester	150–350	400–650
Format	Coordinated sections of 25–35 students	Instructor-led lectures of 150–250 students; TA-led discussions of 22-25 students
Instructional hours per semester	35 class meetings, 75 minutes each	42 lectures, 50 minutes each; 14 discussions, 50 minutes each
MC modules used	All activities in Table 3.	• Intro to multivariable functions • Traces of a saddle • Visualizing partial derivatives • Triple integral • Parametrizing a paraboloid • Stokes' Theorem Example 2
Time spent on MC activities	11 class meetings included MC activities; 5–40 minutes per activity	MC activities were the primary activity in 5 discussion sections; one 10 minute MC activity was used in lecture

3D models in mathematics instruction more broadly. In Section 2, we give background on the rationale for using 3D models in calculus instruction and an overview of the use of 3D models and 3D printing in mathematics education. Sections 3 and 4 describe the process of designing and producing the models and classroom activities. Finally, in Section 5, we share experiences from the implementation of the lessons in the multivariable calculus courses at Harvard and Cornell, which have similar content but very different course structures.

Curricular materials associated with Manipulative Calculus, including source files for 3D printed objects, can be found on the project website [**10**].

2. Background

The Manipulative Calculus project implements research findings on the role of 3D models in classroom environments and builds on other curricula that use 3D printed models to support student learning. In this section, we give an overview of literature about the use of 3D printing and 3D printed models in mathematics classrooms, along with education research supporting the inclusion of 3D models in the teaching and learning of calculus.

Broadly, student engagement with physical manipulatives serves many purposes in STEM education [**29**], and the importance of active learning in this setting has been well documented [**3**, **7**]. While student interaction with 3D models has long been considered an essential component of instruction in fields like chemistry and anatomy, the use of such models is far less ubiquitous in mathematics education. At the same time, student struggles with concepts involving spatial reasoning are well-documented in courses like integral [**18**] and multivariable [**15**] calculus, and research has suggested that spatial skills are an important mediator of student performance across STEM disciplines [**31**].

In education research, the field of semiotic representation theory studies how students use the resources available to them to communicate and build understanding of mathematical concepts [**22**]. These resources may include classroom lecture and discussion, written notes, drawings, and physical models. Depth of knowledge of a particular concept may be associated with a student's ability to coordinate the understanding derived from different types of resources [**6**, **19**]. In the context of calculus instruction, it therefore makes sense to include examples that require coordination of geometric and symbolic representations of the same concept. Physical 3D models as a resource in studying concepts in three dimensions may be particularly beneficial because they most closely represent the underlying mathematical objects [**20**].

In integral calculus, one can find many examples of instruction using 3D models —both 3D printed [**5**] and otherwise [**2**, **8**, **30**]—for teaching topics with inherent spatial components, such as calculating volumes and certain calculations of physical work.

Multivariable calculus courses typically cover many topics related to three-dimensional geometry and are required for many students planning to enter STEM fields. Several authors have proposed using 3D models for instruction of multivariable calculus. The types of models described include ones hand-made by students [**1**, **13**], magnetic pegboard kits for assembling various objects [**16**], thermoformed plastic models [**32**], and 3D printed models [**5**, **21**].

Beyond the calculus sequence, 3D printing and 3D models more generally have been used in mathematics education largely for illustrative purposes [**4**, **26**]. However, notable exceptions exist in mathematics courses where students design and print their own 3D models, such as those listed in Table 2.

The impact of using 3D manipulatives in undergraduate mathematics classrooms is the subject of ongoing research. At the calculus level, positive effects on students' problem-solving achievement [**16**] and confidence [**8**] have been observed.

At the same time, using 3D printed models for classroom instruction has never been easier. Makerspaces with access to 3D printers can now be found on many college and high school campuses and at public libraries. Repositories of freely downloadable 3D object files, typically in the STL (stereolithography) file format, can be found on the web at sites like Thingiverse [**14**], and it is easy to find STL files for models designed for mathematics education. All of the 3D printed objects from the Manipulative Calculus project are available on Thingiverse, and links can be found on the project website [**10**]. For those without access to a 3D printer, 3D printing services can be found on the web, for example, through Shapeways [**28**].

TABLE 2. A sampling of university-level mathematics courses in which students design and print 3D models

Course	University, Instructor
Various courses associated with 3Space [11]	James Madison University, Laura Taalman and others
Mathematical Computing [9]	Queens College, Chris Hanusa
Mathematics Through 3D Printing [24]	George Mason University, Evelyn Sander
Geometry and Algorithms in 3D Modeling [25]	Oklahoma State University, Henry Segerman
Various cross-listed courses in mathematics and art	Pitzer College, David Bachman

For those interested in designing their own models, many commonly used mathematical software applications like Mathematica now support output of STL files. Noteworthy among such software is the open-use web application CalcPlot3D [23], which is specialized for instruction in integral and multivariable calculus.

3. Design and production

In this section we describe the process of designing the lessons and accompanying 3D models that make up the Manipulative Calculus project, and we share experiences of producing the models.

3.1. Design principles. We designed the 3D models and classroom activities in tandem with the following considerations in mind.

Active learning: Our foremost guiding principle was that the lessons should involve student interaction with 3D models. While there may be value to using 3D models for purely illustrative purposes, our focus was on putting models into students' hands and choreographing problem-solving experiences in which interacting with models plays an essential role. Whenever possible, we wanted the classroom activities to be opportunities for students to discover key ideas by themselves.

Connect the geometry to computations: While interacting with 3D models may build students' geometric understanding of course concepts, our lessons are particularly focused on integrating this geometric understanding with analytic and computational techniques. The authors feel that a robust understanding of multivariable calculus necessarily involves coordination of the geometric and analytic viewpoints. Ultimately, students' success in the course will be measured primarily through instruments that focus on analytic skills, but geometric understanding should support those skills and foster confidence and sense-making.

Tailor 3D models to problems: The lessons designed for Manipulative Calculus are based around students solving problems from worksheets. Typically when worksheet problems make reference to the 3D models, students need to collect (through observation) or add (by drawing) information to the model. Because of this, we sought to tailor the information delivered by a model itself to the questions we wanted to ask about it. For example, for a lesson on surface parameterization that uses the paraboloid as an example, one might consider using one of the two models shown in Figures 3A and 3B. In a sense, the wireframe model delivers more information to students, but that may not be desirable if we want to ask students to draw or visualize these grid curves on their own.

Target common misconceptions and difficulties: From our prior experience teaching multivariable calculus, we knew of several points in the course where visualization was a barrier to student understanding, so we tried to design models and activities that would address these. For example, we had observed that students often initially found it difficult to visualize partial derivatives on a contour map and ended up memorizing rules for doing this. So, we wrote an activity focused on having students look at partial derivatives given both the graph and contour map of a function.

Scalability: As a practical matter, we needed to design our 3D models so that we could scale production to meet the needs of the entire multivariable calculus course. At Harvard, multivariable calculus has one of the highest enrollments of any course taught at the university. Assuming students work in groups of approximately three and taking into account scheduling overlaps, this meant being able to produce approximately 60 copies of each model. We also wanted to make the lessons useful for larger universities, where productions runs may be in the hundreds.

Multi-lesson applicability: In multivariable calculus, students learn about several types of analysis that can be done on the same underlying construct. As one example, students learn to analyze surfaces by describing them implicitly and parametrically, by relating surfaces to functions of two and three variables, and by integrating both scalar and vector fields on the surface. For this reason, the same physical 3D model can often be useful in teaching several different lessons, so we looked for opportunities to design models that would have this multi-functionality.

Some topics require consideration of a specialized geometric object that may not be useful elsewhere in the course, such as the topic of multivariable limits and continuity. As a matter of practicality, in these cases we generally made the decision not to produce class sets of physical models.

Practicality: There are various other practical considerations in designing 3D models. Many of these were either anticipated by considering the needs and resources of students and instructors or were discovered through experience while prototyping models and piloting lessons. These considerations include things like durability, making models easily portable, ensuring that models are free-standing, and that models appropriately interface with each other as needed.

Usability: Since Math 21a is taught by a team of instructors with varying levels of experience, it was important to create activities that would be easy for instructors to adopt.

3.2. Pilot workshops. We began the design of the models and lessons in Spring 2018 at Harvard by holding pilot workshops using prototype models. The workshops were offered to all Math 21a students that semester as optional extra practice.

Here's one problem from a pilot workshop; the goal was to give students some hands-on practice with traces (slices of a surface where one variable is held constant).

Workshop sample problem

This problem deals with the surfaces defined by the equations $z = x^2 - y^2$ and $z = 2xy$. Your group will be given a few models of surfaces, and your goal is to figure out which model corresponds to each equation.

(a) Sketch the traces with respect to each of the coordinate variables for each of these two surfaces.

For $z = x^2 - y^2$:

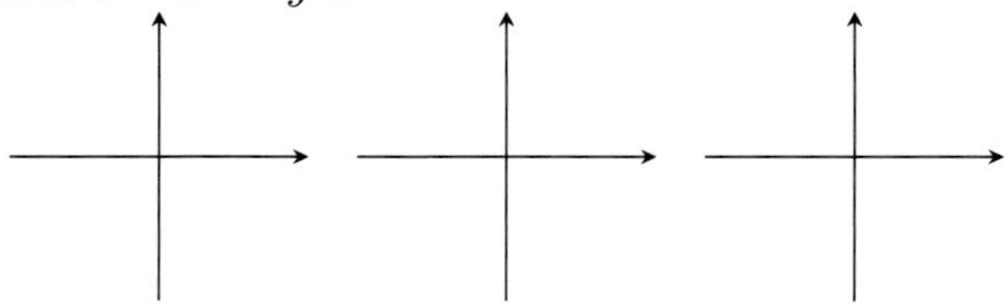

For $z = 2xy$:

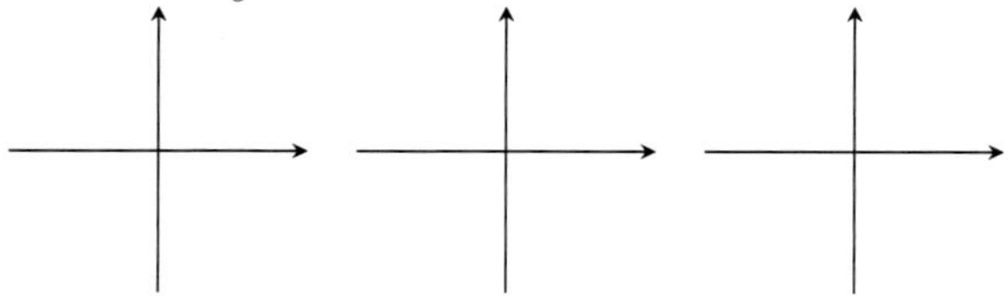

(b) Use the models provided to illustrate how the traces you found can be seen as cross-sections of the surface.

(c) How, specifically, are the two surfaces different? Would it have been possible to see the difference just from the equations and without drawing the pictures?

Watching a group of students work through this problem was very revealing. Although the students were able to figure out which model was described by $z = x^2 - y^2$, they had some gaps in understanding that only became clear when they tried to draw on the model. For example, they had learned in class that the way to calculate traces algebraically is to set one variable to be constant, so they were comfortable saying that the trace in $y = k$ is $z = x^2 - k^2$ and identifying this as a parabola. However, when they tried to draw this parabola on the model, they realized that they didn't understand what k represented. They also imagined the trace in $x = k$ to be an upward parabola and had difficulty reconciling this with the 3D model.

During the workshops, there were multiple times where a student would answer a standard problem correctly, but when they used a model to illustrate their thinking, we would realize they had deep misconceptions. Moreover, some of the misconceptions were ones we had never realized students might have, despite the fact that we had taught multivariable calculus multiple times before. These experiences convinced us that bringing the models into the classroom for all students to use would be beneficial to their learning.

3.3. Design process for the function graph. In this subsection, we do a deep dive into the design of one of the models used in several of the Manipulative Calculus lesson activities to give an example of how both pedagogical and practical considerations informed the design process.

Functions of two variables are one of the central concepts studied in multivariable calculus, and in the course they are typically represented geometrically in two ways. A *contour plot* of a function $f(x,y)$ consists of a collection of loci in the domain $\mathbb{R}^2$ along which the function is constant. The *graph* of a function $f(x,y)$ consists of the points (x,y,z) in $\mathbb{R}^3$ where $z = f(x,y)$. While the contour plot of a function of two variables can be faithfully represented with a two-dimensional drawing, the graph is a surface living in three dimensions, so only a projection of the graph can be shown in a two-dimensional image. Drawing on instructional practices suggested by Wangberg [**32**] and Beckman and Schlicker [**1**], the Manipulative Calculus materials include a model of the graph of a function of two variables, shown in Figure 1A.

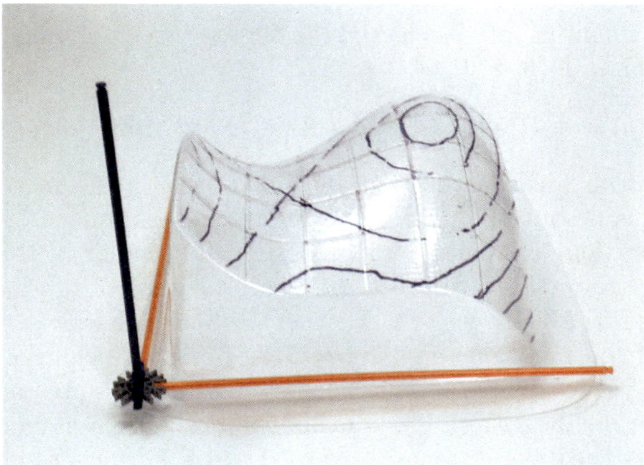

(A) Function graph model shown with contours marked in dry-erase marker and K'Nex axes.

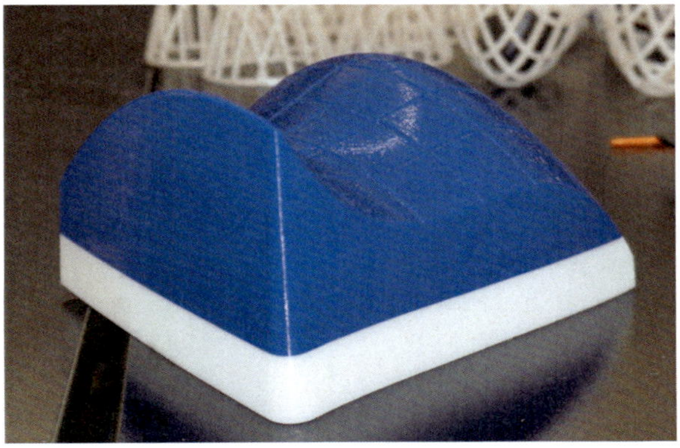

(B) 3D printed buck for thermoformed function graphs.

FIGURE 1. Thermoformed function graph model.

The function graph model we designed draws most directly on the models used in the Raising Calculus to the Surface curriculum [**32**]. However, in order to interface with the existing multivariable calculus curriculum and course structure at Harvard, we opted to design and produce our own models rather than use the existing Raising Calculus models.

One of the first steps in the design process was selecting an appropriate function of two variables. We wanted the function graph models to be an effective learning tool in a variety of lessons relating to functions of two variables including the introduction to functions of two variables, partial and directional derivatives, gradient, and optimization. Some of the important pedagogical considerations in selecting a function were

- the function should have features important to each lesson (e.g. critical points for the optimization lesson) on the represented domain,
- the function should have an explicit formula that is simple enough to be used in calculations, and
- the function should lack symmetries that would allow students to avoid engaging with the concepts directly.

After some experimentation, we selected the degree-three polynomial function given below on the domain $[0,6] \times [0,6]$.

$$f(x,y) = \frac{1}{30}\left(-2x^3 + 18x^2 + 3xy - 3y^2 - 51x + 9y + 107\right)$$

For the optimization lessons, the function has both a local maximum and saddle point in the domain, but the global maximum and minimum of the function on the domain occur along the boundary by design. The range of the function on the domain is smaller than the width of the domain for ease of production.

We wanted to be able to write worksheet problems that refer to specific points in the domain of the function (e.g. "is $f_x(1,3)$ positive, negative, or zero?"), and so we needed to make integral points (ones whose coordinates are both integers) easily identifiable on the graph. To do this, we designed our mold to have subtle indentations along the integer traces in x and y on the surface, essentially overlaying a grid.

Students were given a K'Nex kit with materials to make coordinate axes as shown in Figure 1A so that they could keep track of which corner is at the origin. One advantage of using the K'Nex axes to mark the origin rather than, say, permanently marking the models with axes is that the same model can be used to represent four different functions since it can be rotated. In certain worksheet problems, as a warmup step, students were shown a contour plot of the function and then asked to properly orient the function graph to match.

Since the integral points are easy to identify visually, we were intentional about which features of the function occur at integral points. The critical points, for example, do not occur at integral points so that students cannot determine their coordinates without calculating. However, other features do occur at integral points so that they could easily be referenced in worksheet problems, especially ones where the contour plot is also shown. For example, the level set where $f(x,y) = 2.7$ precisely passes through the integral points $(1,1)$, $(1,3)$, $(4,1)$ and $(4,6)$ in the domain, and this level set has a horizontal tangent at $(4,1)$. The horizontal tangent feature is important for a problem which targets a fact that often confuses students about partial and directional derivatives—that the derivative in a direction tangent to the level set is necessarily zero. See Section 4.3 for details on how this feature is leveraged.

Since the function graph models would be used by groups of three students and would need to be transported from classroom to classroom, we decided that a 6 inch square base would be an appropriate scale. However, 3D printing dozens of models of this size would be slow and expensive, so we instead 3D printed a single copy of solid region below the surface and above the xy-plane, shown in Figure 1B to be used as a mold over which plastic sheets could be thermoformed.

In general, thermoforming as a production method involves heating a sheet of material until it is pliable, and then vacuum-forming the pliable sheet over a mold called a "buck". After experimenting with various plastics, we decided to use 0.04 inch thick clear PETG sheets (think plastic soda bottles) since they formed well during test runs, and the resulting models are durable, transparent, and work well with dry-erase markers. Initial attempts with polystyrene (think disposable coffee cup lids) sheets produced low-quality results.

Designing a 3D printed buck for thermoforming requires several practical considerations. First, the printer material must be heat-resistant enough so that repeated exposure to the heated plastic sheets does not cause warping or other damage to the buck. At Harvard we used ABS plastic (think Legos), and at Cornell, we used PETG; both plastics have relatively high melting points. Bucks made from PLA plastic (think biodegradable plasticware) do not work well with thermoforming; see Figure 2 for a failed attempt with a buck printed with PLA filament.

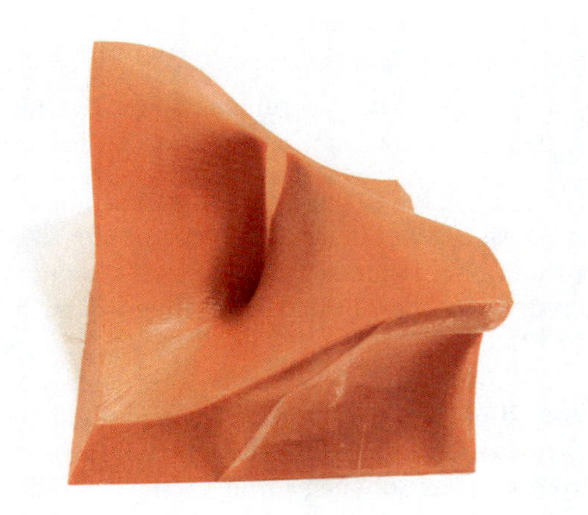

FIGURE 2. Deformed PLA buck. This 3D printed model was intended to be used as a buck for thermoforming, but it was printed in PLA, which has a low melting point. The vertical faces were originally flat, but here we see them badly warped after a single attempted thermoform. Photo credit: Edmund Harriss, University of Arkansas.

Another important consideration is that the buck must be removed from the cooled plastic sheet after each form, so the bucks cannot have vertical or overhanging faces. For this reason, our 3D printed buck was designed with a 3° "draw" on the vertical faces; this also makes the finished models easy to stack. Furthermore, the buck must not have any sharp edges or corners, as these may cause the heated plastic sheet to puncture or tear. For this reason, we designed the buck to have rounded edges and corners.

In the slicing stage of the 3D print design, we used medium-density infill to strike a balance between durability and breathability, both of which are required for use in thermoforming. Also, we intentionally set the top layer of the print to be low-resolution (0.3 mm) so that striations would be visible on the buck and would render on the thermoformed models. The result is that the level sets of the function form a texture that can be seen faintly on the function graph models.

As with all of the 3D printed Manipulative Calculus materials, the STL file for the function graph mold was designed using the script-based design software OpenSCAD [**12**]. While software like Mathematica have the ability to export STL files, most lack the fine-grain control of the output or use computational shortcuts that result in defects in the output file. While OpenSCAD may be unfamiliar to instructors hoping to use and adapt the models to their own classes, there is excellent documentation for the software available and the parameters are easily changed in the script. OpenSCAD also has the virtue of being free.

An interesting peripheral note to make here is that the task of scripting the design of the buck was ultimately an exercise in applying concepts from multivariable calculus. In retrospect, this would have been a good project for an advanced undergraduate student.

3.4. Other design considerations. Pictured in Figure 3 are the 3D printed models that are used in Manipulative Calculus lessons. In this subsection we collect various practical and pedagogical design considerations for these models.

- We found that an effective way to foster student interaction with the models was to assign worksheet problems which asked students to mark features on the models themselves. For this reason, all 3D printed models except the wireframe paraboloid were finished with blackboard paint as suggested in [**21**].
- In order to facilitate group work with the models, we found that it was important for all of the models to be freestanding. The rims of the paraboloid and hyperboloid are designed with this in mind, and small cut-outs are made from the base of the saddle. The parabolic solid in Figure 3D is designed with a hole through the middle so that it will interface with the K'Nex axis kit.
- Paraboloids and saddles are stackable. Hyperboloids are not. This makes a difference for storage and portability.
- To be 3D printable, surfaces must be endowed with some thickness. We found that a thickness of 0.1 inches provided sufficient durability for the paraboloid, hyperboloid, and saddle models. We achieved this by extruding 0.05 inches in both normal directions from the parameterized surface.
- In the slicing phase of the design, we used medium infill on solid models. Only the saddle and parabolic solid required breakaway supports. Several copies of each model could print in the same run, helping us maximize efficiency in production.

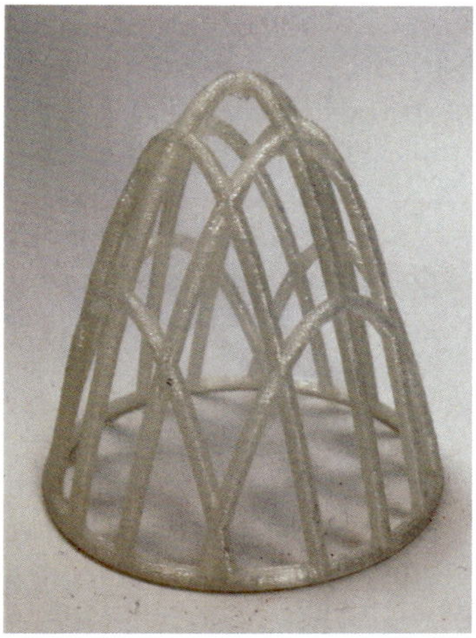

(A) 3D printed wireframe paraboloid.

(B) Paraboloid.

(C) Saddle.

(D) Solid bounded by a paraboloid and parabolic cylinder.

(E) Hyperboloid.

(F) Three solids with rotational symmetry.

FIGURE 3. 3D printed models.

3.5. Limitations. The models in Figures 1 and 3 covered many but not all of the topics we hoped to create manipulatives for. In particular, we knew that students often struggled with visualizing ideas from vector calculus. For example, when first encountering flux integrals, students often find it difficult to separate the roles of the vector field and the surface. We had many discussions about ways to create a model of a vector field on $\mathbb{R}^3$, but we weren't able to arrive at a practical design that students would be able to interact with meaningfully. For topics like this, we rely more on software such as GeoGebra to help students visualize the objects in question.

For practical reasons, we did not produce enough models for students to have their own copies. Thus, students do not have easy access to the models when doing their homework. So, when we want homework assignments to reinforce a visualization students explored in class, we typically ask students to interact with GeoGebra applets. In addition, instructors often complement activities involving manipulatives with in-class demonstrations using dynamic geometry software.

3.6. Production. The first production runs of the models used in Manipulative Calculus happened in the summer of 2018 at Harvard University. Funded by a Course Innovation Grant through the Office of Undergraduate Education at Harvard University, we produced 60 copies of each of the models from our curricular materials—enough to serve the 150–350 students typically enrolled in Math 21a each semester with plenty of spares. We hired a project assistant, Dixon Magister, for the summer to carry out the production in two on-campus makerspaces, the Active Learning Labs and the Instructional Physics Laboratory. Production work involved managing several hundred 3D printer jobs, finishing models with blackboard paint, pressing and trimming the thermoforms, and troubleshooting printer and thermoformer issues. At Harvard, all of the 3D printed models were printed with PLA plastic, and the function graph models were made on a thermoformer that was constructed in-house at the Active Learning Labs makerspace.

Along with the models, we assembled pencil box kits of materials for students to use with the models, one of which is shown in Figure 4. Each kit includes a few sticks of colored chalk and a tiny chalkboard eraser (for writing on the 3D printed models), a few dry-erase markers with eraser caps (for writing on the thermoforms), and a K'Nex axis kit. These kits were handed out to students anytime we did a lesson with one or more of the models.

FIGURE 4. Pencil box kit

A second production run was done at Cornell University in the Fall of 2021. This project was supported by Cornell's Center for Teaching Innovation Micro Grants Program, along with matching funds from the Mathematics Department. We planned to use the models in the Spring of 2022, when enrollments of Math 1920 Multivariable Calculus for Engineers were predicted to be 650+ students. Since models would be used in TA-led recitations with overlapping times, each TA would need their own set of models. Thus 60 copies of each model would be needed.

At Cornell, we chose to manufacture the 3D printed parabolic solid, saddle surface, paraboloid, and the thermoformed function graph. Production at Cornell was carried out by the second author, who prior to this project, had very little experience with 3D printing. For those considering producing their own models, we give a detailed account of how we used the Manipulative Calculus resources for our production.

All of the 3D printed models were printed on a single Prusa i3 MK3s+ using either PLA or PETG plastic. Prusa printers run on gcode that is created by the open source software application PrusaSlicer. This slicer has simple, advanced, and expert settings; for this project the simple settings were sufficient, and proved straightforward to use. For the saddle surface and paraboloid, we downloaded the STL file from thingverse and made no edits; we simply opened the STL file with the Prusa slicer, and scaled the model up to be about 4 inches tall (scale factor approximately 2400%). We choose a different building set for our axes, and therefore needed to make a small adjustment to the size of the hole in the parabolic solid. This required installing OpenSCAD and making one change to a parameter. We note that the second author typically uses Mathematica, and has essentially no experience with OpenSCAD. Despite this fact, from installation to creating the new STL file took about 15 minutes.

FIGURE 5. Producing parabolic solids.

Models were sliced at the .15mm "quality" setting with 15% infill. The entire project required approximately 1000 hours of printing time (4–6.5 hours per model). While the printing time for this project was substantial, the printer ran unattended for the vast majority of this time. Plating multiple models made the production much more efficient. For example, the parabolic solid took 4 hours and 20 minutes to print; however one can print up to 9 models at a time. We preferred plating 6 at a time, which takes about 25 hours. This meant we only had to visit the printer once a day to set up the next print, which took about 15 minutes. Both the paraboloid

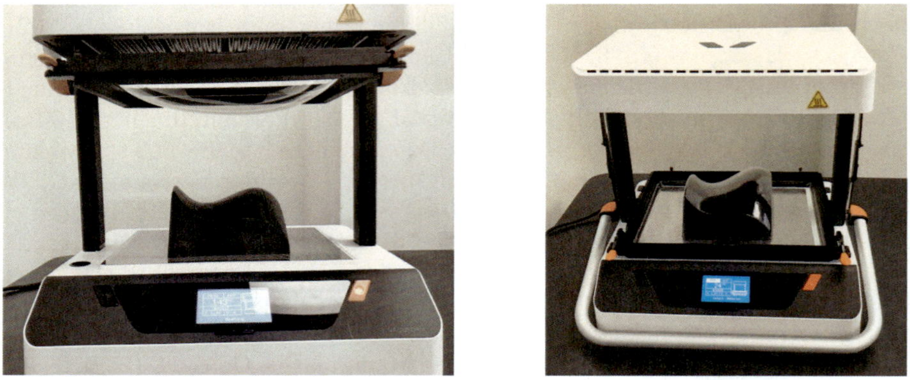

(A) Ready to pull. Notice that the plastic sheet at the top of the machine is bowed by the heater above it.

(B) After vacuum forming. The quality of the pictured model is very good with no "webbing" at the corners.

FIGURE 6. Production at Cornell with a Vaquform thermoformer.

and parabolic solid were very stable prints, with essentially no failures. Plating multiple saddle surfaces led to more failures, so we printed those individually. For a class with enrollment of 25 students, 6 of each model would allow for about four students per group. For these three particular models, this would require about 6 days of printing, but only 1–2 hours of attention.

Finding access to a vacuum thermoformer on campus proved more challenging. To manufacture the function graph, we purchased a Vaquform desktop digital vacuum thermoformer, shown in Figure 6. Unlike 3D printers, for which many options exist, this thermoformer was the only affordable option we could find with a large enough forming bed to create the 6×6 inch function graph model. Fortunately, after an initial setback with a defective part, the thermoformer worked well for this project. We used PETG 1mm plastic sheets which resulted in sturdy models. One advantage of the digital thermoformer is control of the temperature to which the plastic sheet is heated. Optimal forming temperatures depend both on the material used (type of plastic and thickness) and the shape of the buck. For the function graph, we found that heating the sheets to 145°C reliably resulted in little to no webbing at the corners. See Figure 6A for an image of the heated plastic sheet that is almost ready for forming, and Figure 6B for the result of a successful "pull".

As mentioned, our buck was printed in PETG plastic. For this particular model, the PETG buck worked well. However, experiments with other bucks printed in PETG have been less successful. In the future, we plan on printing bucks in ABS plastic.

3.7. Production options for potential adopters. For those interested in using the Manipulative Calculus lessons, we list here some possibilities for producing class sets of the 3D models.

DIY 3D printing: As we mention in Section 2, 3D printing technology has become quite accessible; makerspaces, schools, and libraries often have 3D printers available for public use, and high quality 3D printers are now more affordable than ever. Depending on the resources available to you, your level of expertise, and the size of your production run, it is often most economical to do the 3D printing yourself.

3D printing services: If you have STL files for the objects you'd like to 3D print, you can have a third party like Shapeways [**28**] do the 3D printing for you.

Finishing: If you plan on asking students to write on the models, as many of the Manipulative Calculus lessons do, you'll need to take the extra step of finishing the models. In the implementations used at Harvard and Cornell, the models were finished with blackboard paint so that they would be writable and erasable with chalk. The authors have heard that others have had success making their 3D prints writable and erasable with dry-erase markers using a smoothing finish like XTC-3D resin, made by Smooth-On [**27**]. Third party 3D printing companies may also have finishing options.

Thermoforming: Thermoformers are more specialized and less ubiquitous than 3D printers. A well-equipped makerspace may have one, but many do not. Purchasing a tabletop thermoformer like the Vaquform used in Cornell's production run may be a feasible option for some, but we should mention that 3D printing a class set of the function graph models may be simpler and more economical for smaller production runs. In this case, we recommend a very low infill setting to minimize print time and material cost. We also note that scaling these larger models down to 80% will result in approximately 50% less printer time and materials.

4. Classroom activities

Here, we'll describe four of the classroom activities in detail, to give a sense of some ways we asked students to use the models. The first and third activities use the thermoformed function graph and a set of K'Nex axes. The other two activities use chalk and the 3D printed saddle and paraboloid in Figures 3C and 3B. In Table 3, we list all of the activities available on the project website [**10**], along with a very rough estimate of the time we spent in class on each activity in Fall 2018 and Spring 2019. It should be emphasized that these approximations are very rough; each instructor decided how much time they wanted to spend on an activity, and the chart below only reflects the authors' decisions. Also, we designed most activities so that the later parts weren't essential; the approximations below reflect the time we allocated to the activities in class, with the understanding that not all groups would finish the entire activity.

Some activities are marked "optional", which means that students who worked at a faster pace on that day's material were encouraged to try the activity.

It is important to note that the intention of these activities is that students will attempt most parts of the problems in small groups but that the instructor will lead whole-class discussions on the concepts as well. These discussions may include emphasis of important or tricky points from the worksheets and lecture on the analytic and computational facets of the concept.

4.1. Introduction to multivariable functions. This activity was intended to be done on the first day of the course, so it had several goals:

- We wanted to establish the classroom as a space for active learning and immediately set the expectation that students would be working in groups throughout the semester.

TABLE 3. List of activity modules from Manipulative Calculus.

Activity	**Manipulatives used**	**Approximate time spent in minutes**
Introduction to multivariable functions	function graph model, K'Nex kit, dry-erase markers	30–40
Traces of a saddle	saddle, K'Nex kit, chalk	10
Surface from level sets and a trace	K'Nex kit	optional
Describing solids in cylindrical and spherical coordinates	the three solids in Figure 3F	optional
Visualizing partial derivatives	function graph model, K'Nex kit, dry-erase markers	30
Triple integral	the solid in Figure 3D, K'Nex kit, chalk	10–15
Triple integrals in cylindrical and spherical coordinates	the three solids in Figure 3F	10–15
Tangent line to a graph	function graph model, K'Nex kit, dry-erase markers	5
Parameterize a paraboloid	paraboloid, K'Nex kit, chalk	10
Review problems	paraboloid, K'Nex kit, chalk	50
Local and global extrema	function graph model, K'Nex kit, dry-erase markers	10
Lagrange multipliers	function graph model, K'Nex kit, dry-erase markers	5
Stokes' Theorem Example 1	paraboloid, chalk	5
Stokes' Theorem Example 2	saddle, K'Nex kit, chalk	optional

- In terms of mathematical content, the primary goal was for students to understand the relationship between the graph of a function of two variables and the contour map of that function. We especially wanted students to start thinking about how steepness is conveyed in a contour map, as preparation for thinking about partial derivatives.
- We wanted students to practice interpreting the graph of a function in contexts where the function doesn't simply represent height, as well as to hint at ideas that we would be studying later in the semester.

- Extending the idea of contour map, we wanted to introduce the idea that, when studying objects in $\mathbb{R}^3$, it can be very helpful to look at slices of the object.

Introduction to multivariable functions worksheet problem

Your group will be given a model showing the graph of a function $f(x, y)$ where $0 \leq x \leq 6$ and $0 \leq y \leq 6$. Orient it like this (so the highest of the 4 corners is at $x = 0, y = 0$):

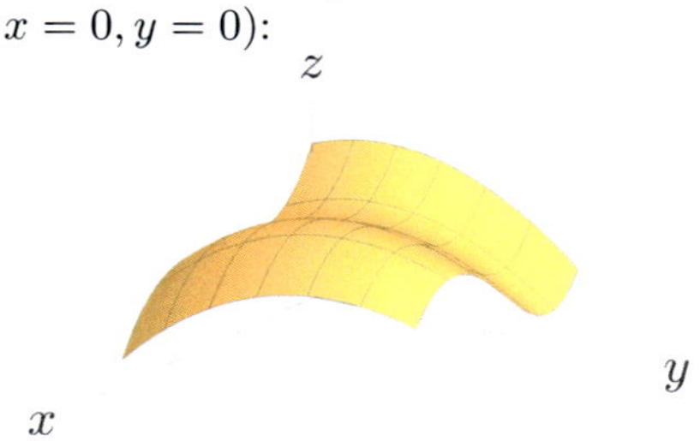

The lowest point has z-coordinate 0.5 and the highest point has z-coordinate 3.5.

(a) Write down the coordinates of a point (x, y) so that $f(x, y) = 3$. Mark the corresponding point on the model.

(b) On the axes below, sketch the set of all points (x, y) where $f(x, y) = 3$. (A very rough sketch is fine.) Also sketch on the model. This set of points in the xy-plane is called a level set of f.

(c) On the same set of axes above, sketch several more level sets of f where $f(x, y)$ is equal to other constants. We call this a contour map of f. Also sketch on the model.

(d) (i) Suppose $f(x, y)$ gives the rainfall in inches at position (x, y), where x and y are measured in miles. Which areas would you expect to be most susceptible to flooding?

(ii) Some places on the graph are "steeper" and others are "flatter". Supposing $f(x, y)$ gives the temperature at position (x, y), describe in a sentence the significance of the steeper parts of the graph.

(iii) If $f(x, y)$ gives the population density in people per square mile at position (x, y), where x and y are measured in miles, approximately how many people live in the region where $1 \leq x \leq 3$ and $1 \leq y \leq 3$?

(iv) Suppose $f(x, y)$ gives the number of cars per km in hundreds at position x along a highway at time y. Is the traffic at position $x = 5$ getting better or worse with time when $y = 2$?

(e) (i) On your model, sketch the intersection of $z = f(x, y)$ with $x = 5$. (We call this the trace of $z = f(x, y)$ in $x = 5$.) Then, sketch the same trace on the axes below (since x is constant, we ignore it).

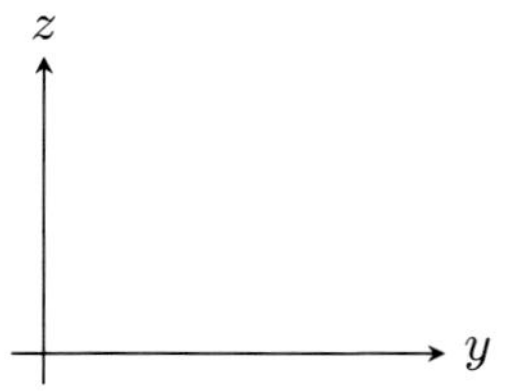

(ii) Sketch the trace of $z = f(x, y)$ in $y = 2$ on the axes below.

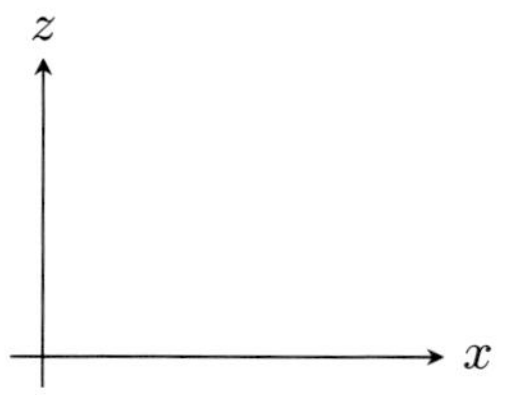

(iii) Which of the two pictures you just drew would be helpful in explaining your answer to part (d)iv?

(f) There's a point on the model that looks like the top of a hill near the point where $x = 4.5$ and $y = 3.8$. How does the contour map look near this point? If you were given the contour map of a different function (but not its graph), how could you recognize the locations of "peaks" like this? How about "pits" (something that looks like the bottom of a hole someone dug)?

(g) Suppose $f(x, y)$ gives the elevation of some landscape at (x, y).

(i) Pick a point where you would describe the landscape as "steep". Sketch a trail, both on the model and contour map, that passes through this point and climbs as steeply as possible. Then sketch a different trail which passes through the same point but climbs less steeply.

(ii) On your contour map, how can you tell the difference between a path that changes elevation steeply vs. one that changes elevation gradually?

4.2. Traces of a saddle. Like the previous activity, we used this on the first day of the course. The goal was to reinforce the idea that we can understand the graph of a function by looking at the contour map and slices of the graph in other directions, as well as to have students start thinking about how to get information from an algebraic formula for a function. This problem is a revised version of the pilot workshop problem discussed in Section 3.2, rewritten to address the misconceptions we saw students have during the workshop.

Traces of a saddle worksheet problem

Now, let's see how we can apply some of the same ideas when we have a formula for a function. In this problem, we'll look at $f(x, y) = x^2 - y^2$.

(a) Sketch the contour map of $f(x, y) = x^2 - y^2$. Please label the value of the function on each level set you draw.

(b) What is the trace of the graph of f in $x = 0$? $x = 1$? $x = -1$? In general, what is the trace in $x = k$ where k is a constant?

(c) You've been given a model of the graph of f. (Do you see why we call the shape of this graph a saddle?) Use your answers to (a) and (b) to figure out how this model should be oriented relative to the coordinate axes. Then, on the model, sketch some of the traces in $z = k$ and $x = k$ that you found in (a) and (b).

(d) Without doing any calculations and just looking at the model, describe the traces of $z = f(x, y)$ in $y = k$. (What is the shape? How does it depend on k?) Sketch some of these traces on the model.

4.3. Partial derivatives. The activity below was used on the day we covered partial derivatives. We first introduced partial derivatives to students in a fairly informal way: to calculate $f_x(x, y)$, think of f as a function of just x, with y being constant; then, take the derivative.

The goal of the activity was to have students figure out how to visualize first- and second-order partial derivatives of a function f of two variables on both the graph of f and the contour map of f.

Partial derivatives worksheet problem

Here's a contour map of a function $f(x, y)$ of 2 variables, shown for $0 \le x \le 6$ and $0 \le y \le 6$.

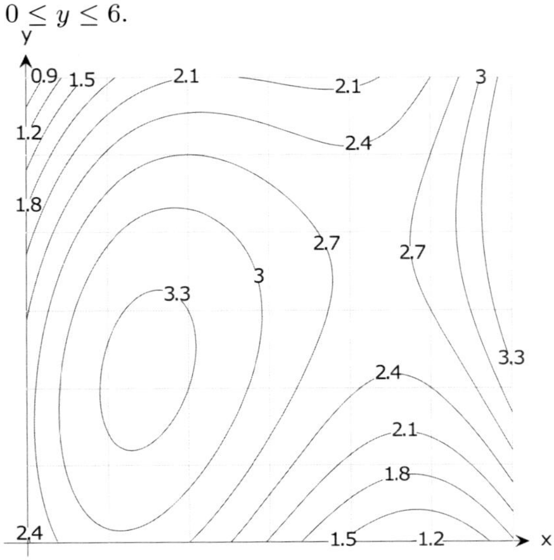

Your group will also be given the graph of f. First, figure out how the graph should be oriented relative to your axes to match the contour map.

(a) On the graph, how do you visualize $f_x(1,5)$? Based on that, is it positive or negative?
(b) How can you see the sign of $f_x(1,5)$ in the contour map of f? Use the contour map to estimate $f_x(1,5)$.
(c) Use the contour map to decide whether $f_y(5,3)$ is positive, negative, or 0. Then use the graph of f to confirm your answer.
(d) Use the contour map to decide whether $f_x(2,5)$ is positive, negative, or 0. Then use the graph of f to confirm your answer.
(e) Use the contour map to decide whether $f_y(1,4)$ or $f_y(1,5)$ is greater. Then use the graph of f to confirm your answer.

Here are two notations for 2nd derivatives: we can write $\frac{\partial}{\partial x}\left(\frac{\partial f}{\partial y}\right) = (f_y)_x$ *as either* $\frac{\partial^2 f}{\partial x \partial y}$ *or* f_{yx}. *(Notice that the order of* x *and* y *depends on which notation you're using!)*

(f) Is $f_{yy}(1,4.5)$ positive or negative? How can you tell from the graph? From the contour map?
(g) Use the contour map to decide whether $f_{xx}(5.5,5)$ is positive or negative. Then use the graph to confirm.
(h) Using the contour map, decide whether $f_{yx}(5,1)$ is positive or negative. How is this reflected in the graph of f?

In part (a), we wanted students to figure out how to visualize a partial derivative like $f_x(1,5)$ as a slope on the graph of f. It has been noted in the literature that making explicit what is meant by "slope" in this context can lead to clearer understanding of this topic [**17**]. When students described such derivatives as slopes, we encouraged them to use K'Nex rods or pencils to illustrate the lines they were talking about. In part (b), we hoped students would be able to connect this to whether f was increasing in a particular direction and figure out how to view this on a contour map. Part (c) addresses a common error; when students look at the contour map, since $(5,3)$ is on the contour where $f(x,y) = 2.7$, students will often move in the positive y-direction from there, see that the next contour they hit is still the contour where $f(x,y) = 2.7$, and want to say that $f_y(5,3) = 0$. When they check this with the graph of f, they should realize their mistake.

Part (d) addresses a similar but more subtle mistake: students often look at the contour map and notice that, if they start at $(2,5)$ and move in the positive x-direction, f decreases, so they'll conclude that $f_x(2,5) < 0$. Again, looking at the graph of f should help students realize their mistake. This idea is particularly important in Math 21a because we build on it; in a later activity, students similarly see that, if $\vec{u}$ is a direction tangent to the level set of f at a point P, then the directional derivative $D_{\vec{u}}f(P)$ is 0. This fact is also important in understanding the key property that the gradient of f is perpendicular to level sets of f.

Part (e) prompts students to recognize that the spacing between contours tells them about the magnitude of a partial derivative, to prepare them for visualizing second-order partials on the contour map. The remaining parts then ask students to visualize second-order partials.

Other subtopics in the lesson on partial derivatives such as interpretation in applications, calculation from formulas, and the symmetry of second-order partials are taught without the use of 3D models.

4.4. Surface parameterization. This activity gives an example of how a 3D printed surface and chalk are incorporated into a lesson. This shorter activity is meant to be done as an early example of surface parameterization, and it can be used to introduce the idea of a parameter domain. We also include the instructor notes for this activity as an example.

Surface parameterization worksheet problem

(1) Parameterize the paraboloid $z = x^2 + y^2 + 1$. On your model of the paraboloid, sketch the grid curves defined by your parameterization.
(2) If we only want to parameterize the part of the elliptic paraboloid under the plane $z = 17$, what restrictions would you place on the parameters you used in 1? Sketch the parameter domain.

Instructor notes for surface parameterization activity

- There are several good ways to parameterize the paraboloid (using x, y as parameters, using r, θ as parameters, or using z, θ as parameters, for example), and if you let students work in groups, you will often see different groups use different techniques.
- Encourage students to draw their two sets of grid curves in two different colors. You can then ask which color corresponds to fixing which variable.
- Students very often tend to draw only one or two grid curves in each direction for some reason. In case you don't get any good drawings, it can be smart to prepare your own models with grid curves drawn on.
- It is worth letting students know what parameterization is useful for, both in direct applications (e.g. computer graphics and design) and later in the class (scalar surface integrals and flux). In particular, it helps to note that the parameterization can be used to dice the surface up into smaller (approximately parallelogram-shaped) bits, which is useful for integration. Similarly, we need to know the bounds on the parameters (i.e. the parameter domain) to set up these integrals.
- One can easily substitute the saddle or hyperboloid for this activity.

5. Experiences and assessment

In this section, we describe our experiences implementing the activities at Harvard and Cornell, including some practical considerations. We also share some feedback gathered from instructors and students about their experiences with Manipulative Calculus.

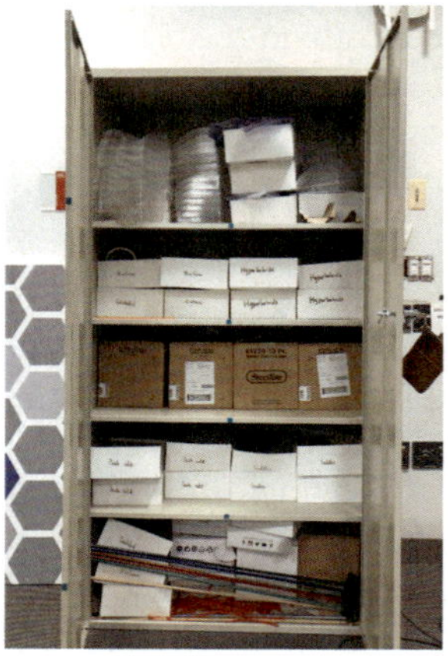

FIGURE 7. Storage cabinet

5.1. Logistics. Harvard's most popular multivariable calculus course, Math 21a, enrolls 150–350 students per semester and is taught in sections of 25–35 students by a team of instructors. The sections are closely coordinated, with all students doing the same homework and exams. The content goals for each class meeting are also closely coordinated, but individual instructors choose how best to achieve those goals in their sections.

In the production run at Harvard, we produced enough copies of the models to have multiple sections use them at the same time, but we didn't have enough for each instructor to have their own set during the semester. Therefore, we made a schedule for how instructors would transfer their models to each other throughout the day. In addition, we needed a place to store the models between classes. The math department bought a storage cabinet, shown in Figure 7, and gave all of the instructors keys. This meant that instructors were also able to access the models during office hours, which was often helpful when asking students to explain their thinking.

5.2. Training. Because Harvard's Math 21a is a multi-section course with several instructors, providing support and training was important. Each section has an instructor (who can be a graduate student or faculty member), as well as 1–3 undergraduate course assistants who facilitate in-class activities. We encouraged the instructors to use the classroom activities but didn't require it. Below are some of the ways in which we supported instructors using the materials.

- One author taught a practice version of the first class where all of the instructors participated as students. This helped the instructors think about possible student misconceptions they might encounter, as well as the basic logistics of distributing models to their class and having students clean the models after they were done drawing on them.
- We wrote teaching notes for the activities and held weekly meetings with the instructors to talk through upcoming activities. See Section 4.4 for an example.

- We also held weekly meetings with the course assistants to familiarize them with upcoming activities and talk about student misconceptions to look for.

5.3. Feedback. In Fall 2018 and Spring 2019 at Harvard, we gathered feedback from instructors and students about their experiences with Manipulative Calculus.

Most of the feedback we gathered from instructors came from our weekly instructor meetings and informal conversations. Instructors who had taught the course before without models felt that the models enabled students to discover more ideas on their own. For example, when we had taught about partial derivatives in previous semesters, students typically weren't able to figure out how to visualize mixed partials on their own and needed to be given direct instruction. Using the models, many more students were able to figure out on their own how to visualize mixed partials.

On the other hand, some props we brought into the classroom didn't appear to have much impact. For example, vector geometry is a challenging topic for students, and we have students do problems on finding the distance between objects (such as between a point and a line, or a point and a plane), as a way to practice applying the vector tools they've learned. We gave students K'Nex rods and sheets of cardboard to use as lines and planes, but students were reluctant to use these to illustrate their thinking unless specifically prompted to by an instructor. Students interacted much more with the 3D models we designed.

We used anonymous surveys to gather feedback from our undergraduate course assistants and students.[1]

Since the undergraduate course assistants we had in 2018–2019 had mostly taken multivariable calculus without models, we asked them whether they thought they would have learned multivariable calculus better as students if they had used 3D props in class. 50% of those surveyed said yes, 36% maybe, and 14% no.

We also included questions about using the models in the students' anonymous end-of-semester evaluations. The results of one such question are found in Table 4.

TABLE 4. Summary of student responses to the question "There were concepts I understood better because of the 3D models used in class. (1=strongly disagree; 2=disagree; 3=neither agree nor disagree; 4=agree; 5=strongly agree)".

	1	2	3	4	5	Mean
Fall 2018	5	17	39	76	94	4.03
Spring 2019	1	4	6	43	59	4.37

We also asked students a free response question about which topics they found the models most helpful for, the results of which are found in Table 5.

From the following student responses, we see that some students found the models particularly useful:

[1]Since this data was not collected for research purposes, the surveys were not reviewed by our Institutional Review Board.

TABLE 5. Most common responses to the prompt "Tell us about one time that you found working with the 3D models helpful."

Topic	Number of students who mentioned this topic in their response	
	Fall 2018	Spring 2019
Partial and/or directional derivatives	51	26
Visualizing objects such as 3D graphs, slices of surfaces, and intersections of surfaces	37	27
Contour maps or level sets	21	7
Parametric surfaces (particularly grid curves)	21	7
Total number of responses	*165*	*81*

- Understanding single and double partial derivatives would have been so much more difficult without the models. Being able to take my pencil and push it along a path and see how the slope changed gave me such a strong conceptual understanding.
- During parameterization, I was really able to visualize the parameters that I had chosen when we drew in chalk on the 3D printed models. I honestly pictured the 3d models each time we were asked to visualize something because they were so helpful in understanding parameters and orientation.

5.4. Implementation at Cornell. At Cornell, Math 1920 Multivariable Calculus for Engineers has a reputation among students as being the most challenging of the four introductory math courses for engineers. In a given semester this course has 400–650 students enrolled, and is taught in large (typically 200+ students) lectures. Students in Math 1920 are also required to enroll in a once a week 50 minute recitation section with at most 25 students. These discussion sections are led by a graduate student TA who is assigned three sections.

In an optional pre-semester survey, Spring 2022 Math 1920 students were asked "Is there anything in particular you were hoping to learn about in Math 1920?" and also "Do you currently have any concerns about math 1920?" Interest in and concern about the 3D geometry in the course was expressed by many students. Below are just a few responses.

"What are you hoping to learn?"

- I'm super excited to learn about Calculus but in the 3D realm—it's super fascinating!
- I'd really like to get better at visualization (especially with 3 dimensions, because I often have trouble dealing with problems that require visualization).
- Perhaps a better understanding of 3D models and graphs.
- a thorough understanding of 3D computations

"Do you currently have any concerns about Math 1920?"

- I heard it's really hard to visualize things in 3D. I haven't had a lot of experiences working with 3D stuffs in Math before so I'm really worried
- I've heard it is the most difficult calculus course to wrap your head around.

- I am concerned about it being too abstract for me to grasp.
- Geometry part of the course
- I am super bad at geometry and did not learn it in high school, so that will definitely be a struggle.

Success in the course requires students to develop a new set of reasoning skills, many of which depend on understanding three-dimensional geometry. We were interested in adopting Manipulative Calculus activities as the activities were designed to support learning such concepts.

Historically at Cornell recitations always feature some active learning; individual TAs create worksheets with some review problems that students work on in groups. Some weeks there is a specified course wide activity called a "workshop", which has more involved problems with an applied focus. On these days, an undergraduate course assistant also attends the discussion to help facilitate the activity. One challenge with the workshops in the past is that students and TAs felt they were somewhat disconnected from the course. Therefore, in looking to adapt the Manipulative Calculus project for use at Cornell, our goals were as follows.

- choose activities that support 3D visualization skills and encourage productive mathematical discussions in recitations,
- ensure activities are well-integrated into the course, and that the students and TAs view them as relevant and engaging, and
- use materials that are relatively easy for (often brand new) TAs to use.

In the end, we chose six of the activities from the Manipulative Calculus Project to use in Math 1920 (see Table 1). The course coordinator can set the expectations for the course, so all TAs are expected to do the scheduled activity. Some of the lessons were slightly altered to make the vocabulary and notation consistent with the course text. We created a sufficient number of models so that the activities done in recitation typically had no more than four students per group. The first semester of implementation is currently underway as of this writing, but initial informal surveys [2] of the TAs suggest that the activities are engaging and encouraged student questions. Students appeared to enjoy working with the models and using the models helped reveal student misconceptions.

We have been primarily relying on the instructor notes created by the Harvard team (see Section 4.4) to give guidance on facilitating the activities. When surveyed, all responding TAs agreed or strongly agreed that the instructor notes "clarified the activity goals" and "gave useful advice for facilitating". We note that none of the responding TAs chose the response option, "didn't read the notes", which we credit to the fact that the notes are concise.

While we have not yet had a chance to survey the students about their experience using the models in recitations at Cornell, anecdotal reports from TAs are positive. After completing the triple integral activity, one TA reported students commenting on how much easier it was to set up the integrals when they had the shape in front of them, and also questioning how the material was taught before 3D printing.

As somewhat of an experiment, we did the surface parameterization activity (presented in Section 4.4) in two lectures with enrollments of around 200 students. We distributed 60 models and chalk to students prior to the lecture starting, and

[2] The surveys were anonymous surveys of the TAs administered on Qualtrics.

had students erase and return them after class (all models were returned). Overall, the activity was a success. Even though the lecture hall had auditorium seating, students were able to work in groups of about three to sketch grid curves on their paraboloids. We found that having the models allowed for better communication between the instructor and students. In both lectures a couple of groups had written down rectangular coordinates for their parameterization of the paraboloid, yet had sketched grid lines corresponding to a polar parameterization. Discovering this led to a productive class discussion that would not have occurred without the models.

6. Conclusion

In this paper we report on the Manipulative Calculus curriculum development project for multivariable calculus. We present samples of the curricular materials, which include digitally fabricated three-dimensional models and lessons designed around student interactions with these models. We share design considerations, both pedagogical and pragmatic, for 3D printing physical models for calculus classrooms. These curricular materials are now being used at Harvard and Cornell Universities–schools whose multivariable calculus courses have different sizes and structures–and we report experiences in producing the models and implementing the lessons at both schools. Finally, we share feedback from students and instructors on the curricular materials.

Worksheets and STL files associated with Manipulative Calculus are publicly available on the project website [**10**]. Please email the authors to request access to instructor notes, solutions, and source files.

Acknowledgments

The authors acknowledge indispensable support from Steven Cortesa of the Harvard University Active Learning Lab and Robert Hart of the Harvard University Instructional Physics Lab in the design and prototyping of the models and the work of Project Assistant Dishen "Dixon Magister" Yang, who was responsible for carrying out the bulk of the production runs at Harvard.

References

[1] Charlene E Beckmann and Steven J Schlicker, *A "model" multivariable calculus course*, PRIMUS **9** (1999), no. 3, 226–240.

[2] Kavita Bhatia, Kirthi Premadasa, and Paul Martin, *Teaching integration applications using manipulatives*, PRIMUS **24** (2014), no. 4, 335–346.

[3] David Bressoud and Chris Rasmussen, *Seven characteristics of successful calculus programs*, Notices of the AMS **62** (2015), no. 2, 144–146.

[4] Diana Davis (ed.), *Illustrating mathematics*, American Mathematical Society, Providence, RI, [2020] ©2020. MR4249620

[5] Elizabeth Denne, *Mathematical Visualization*, `https://denne.academic.wlu.edu/mathematical-visualization/#Calculus`, Accessed Mar 26, 2022.

[6] Raymond Duval, *A cognitive analysis of problems of comprehension in a learning of mathematics*, Educational studies in mathematics **61** (2006), no. 1, 103–131.

[7] Scott Freeman, Sarah L Eddy, Miles McDonough, Michelle K Smith, Nnadozie Okoroafor, Hannah Jordt, and Mary Pat Wenderoth, *Active learning increases student performance in science, engineering, and mathematics*, Proceedings of the national academy of sciences **111** (2014), no. 23, 8410–8415.

[8] Dusty Grundmeier, Stepan Paul, and Deborah Moore-Russo, *3D manipulatives in integral calculus: Student achievement and confidence in solids of revolution tasks*, Manuscript submitted for publication, 2022.

[9] Chris Hanusa, *Mathematical Computing*, http://qcpages.qc.cuny.edu/~chanusa/courses/250/21/index.html, Accessed October 4, 2023.
[10] Harvard University, *Manipulative calculus*, https://www.manipulativecalculus.com, Accessed Mar 26, 2022.
[11] James Madison University, *JMU 3Space classroom*, https://sites.lib.jmu.edu/3space, Accessed Mar 26, 2022.
[12] Marius Kintel, *OpenSCAD*, http://openscad.org, Accessed Mar 26, 2022.
[13] Lorelei Koss, *Manipulatives for 3-dimensional coordinate systems*, PRIMUS **21** (2011), no. 4, 364–376.
[14] MakerBot, *Thingiverse*, https://www.thingiverse.com, Accessed Mar 26, 2022.
[15] Rafael Martínez-Planell and Maria Trigueros, *Multivariable calculus results in different countries*, ZDM–Mathematics Education **53** (2021), no. 3, 695–707.
[16] Daniel McGee Jr, Deborah Moore-Russo, Dennis Ebersole, David O Lomen, and Maider Marin Quintero, *Visualizing three-dimensional calculus concepts: The study of a manipulative's effectiveness*, PRIMUS **22** (2012), no. 4, 265–283.
[17] Daniel McGee Jr, Deborah Moore-Russo, and Rafael Martinez-Planell, *Making implicit multivariable calculus representations explicit: A clinical study*, PRIMUS **25** (2015), no. 6, 529–541.
[18] Batseba Mofolo-Mbokane, Johann Engelbrecht, and Ansie Harding, *Learning difficulties with solids of revolution: classroom observations*, International Journal of Mathematical Education in Science and Technology **44** (2013), no. 7, 1065–1080.
[19] Deborah Moore-Russo and Janine M Viglietti, *Using the K5 connected cognition diagram to analyze teachers' communication and understanding of regions in three-dimensional space*, The Journal of Mathematical Behavior **31** (2012), no. 2, 235–251.
[20] Bernard Parzysz, *"knowing" vs "seeing". Problems of the plane representation of space geometry figures*, Educational studies in mathematics **19** (1988), no. 1, 79–92.
[21] Stepan Paul, *3D printed manipulatives in a multivariable calculus classroom*, PRIMUS **28** (2018), no. 9, 821–834.
[22] Luis Radford, *The seen, the spoken and the written: A semiotic approach to the problem of objectification of mathematical knowledge*, For the learning of mathematics **22** (2002), no. 2, 14–23.
[23] Paul Seeburger, *CalcPlot3D*, https://c3d.libretexts.org/CalcPlot3D/index.html, Accessed Mar 26, 2022.
[24] Evelyn Sander, *Mathematics Through 3D Printing*, https://math.gmu.edu/~sander/EvelynSite/mathematics-through-3d.html, Accessed October 4, 2023.
[25] Henry Segerman, *Geometry and Algorithms in 3D Modeling*, https://math.okstate.edu/people/segerman/geom_algors_3d_modelling.html, Accessed October 4, 2023.
[26] Henry Segerman, *Visualizing mathematics with 3D printing*, Johns Hopkins University Press, Baltimore, MD, 2016. MR3560657
[27] Smooth-On, *XTC-3D High Performance 3D Print Coating*, https://www.smooth-on.com/product-line/xtc-3d, Accessed October 4, 2023.
[28] Shapeways, *Shapeways: 3D printing service*, https://www.shapeways.com, Accessed Mar 30, 2022.
[29] Kok-Sing Tang, Fredrik Jeppsson, Kristina Danielsson, and Ewa Bergh Nestlog, *Affordances of physical objects as a material mode of representation: A social semiotics perspective of hands-on meaning-making*, International Journal of Science Education (2022), 1–22.
[30] Jean Uhl, Patricia Humphrey, and James Braselton, *Visualizing the method of finding volumes by cross sections: An eggsperiment*, PRIMUS **16** (2006), no. 2, 138–146.
[31] Jonathan Wai, David Lubinski, and Camilla P Benbow, *Spatial ability for STEM domains: Aligning over 50 years of cumulative psychological knowledge solidifies its importance*, Journal of educational Psychology **101** (2009), no. 4, 817.
[32] Aaron Wangberg and Ben Johnson, *Discovering calculus on the surface*, PRIMUS **23** (2013), no. 7, 627–639.

Department of Mathematics, Harvard University, Cambridge, Massachusetts 02478
Email address: jjchen@math.harvard.edu

Department of Mathematics, Cornell University, Ithaca, New York 14853
Email address: kd288@cornell.edu

Department of Mathematics, North Carolina State University, Raleigh, North Carolina 27695
Email address: sspaul2@ncsu.edu

Proceedings of Symposia in Applied Mathematics
Volume **79**, 2023
https://doi.org/10.1090/psapm/079/00715

Encouraging student creativity in mathematics through 3D design and 3D printing

Christopher R. H. Hanusa

ABSTRACT. This is a case study of teaching 3D design and 3D printing in a project-based computing course for undergraduate math majors. This article discusses content organization, implementation, project grading, and includes a personal reflection. There is an emphasis on lessons learned and how to encourage student creativity and artistic expression. An appendix details 3D design techniques in Mathematica.

Introduction

As mathematicians we recognize that the research process can be messy and non-linear, that productive failure is a key part of achieving a new result, that creativity and pushing boundaries is necessary to our craft, and that there is beauty in the elegance of mathematical results and visualizations. Now ask yourself: How many of our courses let students experience these aspects that we find so fundamental in mathematics?

This article describes the project-based Mathematical Computing course at Queens College that aims to give students this experience. In this course, students learn the basics of programming in Wolfram Mathematica and actively apply their skills to design mathematical sculpture and create an interactive app. Section 1 describes the structure of the course, including learning objectives and ensuring student success. Section 2 details the philosophy behind and the implementation of the 3D design and 3D printing module in the course, including how students learn about design thinking and the elements of art. Section 3 discusses the written lab report and the way the project deliverables are graded. Section 4 is a personal reflection on the course. Finally, Appendix A summarizes key techniques for 3D design in Mathematica.

2020 *Mathematics Subject Classification.* Primary 97D40, 97N80; Secondary 00A66, 97D60, 97M80, 97P50, 97U70.

Key words and phrases. Wolfram Mathematica, 3D printing, 3D modeling, 3D design, student creativity, design thinking, teaching mathematics, mathematical art, mathematical sculpture, standards-based grading.

The author was supported in part by NSF Grant DUE-1928565.

1. Course structure

1.1. Student population. Queens College is located in its namesake borough of Queens in New York City and is one of the 25 campuses of the City University of New York. The student body of Queens College of approximately 18,000 mirrors the population of Queens, the most ethnically diverse county in the United States. Among freshmen, 44% are immigrants born outside the mainland US, and 34% are first generation students. Queens College serves as an engine of upward mobility and is proud to offer a high-quality education to students at a fraction of the cost of private institutions.

FIGURE 1. Jenny Xu with her 3D printed ceramic pot.

The Queens College Department of Mathematics has 31 full-time faculty members and many part-time instructors. There are approximately 450 math majors in five specialties: pure mathematics, applied mathematics, data science and statistics, secondary education, and elementary education. The largest cohorts are applied mathematics and secondary education.

The course Mathematical Computing (MATH 250) is offered once per year with 75-minute classes twice a week during a 14-week semester. The course has an enrollment between 15 and 25 students, most of whom are juniors or seniors. This course fulfills the computing requirement for applied mathematics majors and serves as a math elective for all other math majors. The prerequisite for the course is either multivariable calculus or linear algebra, to ensure that students have prior experience in multi-dimensional thinking and visualization. The course has no computer programming prerequisite in order to serve the largest possible audience. At the same time, a handful of enrolled students do have extensive programming experience as they are majoring or minoring in computer science.

FIGURE 2. The Fall 2018 Mathematical Computing class.

1.2. Course materials. The current structure of the course was initiated in Spring 2015. There is no textbook for this course; the instructor provides freely available Mathematica tutorials and accompanying video lectures on his webpage.[1] The students contribute to an online discussion platform during the semester for pre-class activities; past platforms have included Blackboard, Microsoft Teams, and Campuswire.

The City University of New York has a university-wide site license for Wolfram Mathematica, which means all Queens College campus computers have Mathematica installed, and all students may install Mathematica on their home machines, so software access and cost are not barriers for our students.

Students have access to the Queens College Makerspace, a new space located in the library that houses approximately eight 3D printers, a 3D scanner, a laser cutter, an embroidery machine, a CNC machine, pen plotters, electronic supplies, and a variety of workshop tools. Once students have completed the orientation, they may freely use any of the materials and machines.

Students are required to pay to print their final model through a third-party 3D printing company such as Shapeways. This is a cost of approximately \$20-\$60 dollars, depending on the size of the model and the material they choose, and is their only expense for the course.

1.3. Learning objectives. Our department believes that every math major should gain some programming experience. For the vast majority of students, this class is their first exposure to any mathematical software. So the most fundamental learning objective is that students learn the basics of programming in Mathematica, developing fluency with the key data structures of lists and functions. Students learn how to design, code, run, test, and debug computer programs. Along the way, students are developing good programming techniques, including sectioning and documenting their code. Students also gain skills to become confident and self-sufficient learners through the use of Mathematica's extensive documentation and internet searches.

Another important objective is that students develop mathematical, programming, and problem-solving skills. For example, students gain a deeper understanding of three-dimensional geometry, including coordinate systems, multivariable functions, and three-dimensional objects. From a programming perspective, students learn and apply functional programming, which is new even for experienced programmers. Building on this knowledge, students apply basic problem-solving skills including analyzing problems, modeling a problem as a system of objects, creating algorithms, and implementing solutions in Mathematica.

A final objective is for students to use mathematics and programming for experimentation, as creative tools, and collaboratively. They apply the design process and communicate the decisions made therein, including ideation, artistic principles, prototyping, and revisions. Students advance teamwork skills by collaborating with classmates, discussing and solving problems in a group setting, and practicing giving and receiving constructive feedback. These ideas will be developed further in Section 2.

1.4. A day in the life. As a project-based course, class time is balanced between content acquisition and project work. The semester breaks down into three modules, each of which culminate with an individual project.

[1] `http://qc.edu/~chanusa/courses/250/21/`

At the beginning of each module, classes are centered around learning the course content. Before each class, students work through a tutorial notebook with an accompanying video lecture, created by the author. Students are encouraged to ask questions about the content through the online discussion board. (A tutorial structure works well given the disparate levels of programming background because the tutorials can be consumed at each student's individual speed.) During class time, students work together in groups on comprehension and challenge questions that reinforce key ideas from the day's tutorial and lecture. During this time the instructor travels from group to group, addressing any questions and, if necessary, gently nudging students in the correct direction. Each class ends with a debriefing of the different ways to approach solving the questions.

As each module progresses, class time turns to project work. Students work on their projects outside class and bring their progress to class for instructor and peer consultation and feedback. As the module nears its end, the instructor spends much of the class period troubleshooting and debugging code and ensuring students are on track to complete their project by the deadlines. Once the students finish their final draft, they complete a peer review with one or more classmates and revise their work before submitting their work for grading.

FIGURE 3. Students working in groups on their projects.

1.5. Course content overview. The first module of the class teaches students the basics of working with Mathematica, exposes students to the wide capabilities of Mathematica, and helps students develop fluency with lists, list manipulation, and functions. The first project involves creating a tutorial centered around a coherent set of Mathematica commands that each student chooses. The instructor encourages students to choose a topic related to the mathematics they have seen in previous courses or from one of the many subject domains to which Mathematica applies. Past student projects include visualizing graphs of multivariable functions, matrix operations, statistical methods, image processing, machine learning, finance, geography, and sound.

The second module of the class is centered around two- and three-dimensional graphics, 3D design, and the 3D printing process. Students design and 3D print a mathematical sculpture. This is the focus of the remainder of this article.

In the third module of the class students learn about and apply the interactive capabilities of Mathematica. The third project involves creating an interactive program similar to an app you might use on your phone. Students determine and design the entire interface and program the behavior of the interface depending on the user input. The resulting programs always vary widely and use a broad spectrum of Mathematica's visual, audio, and interactive capabilities. Past student projects include an interactive map, a piano simulator, a fractal explorer, trivia guessing games, two-player games like dots and boxes or tic-tac-toe, and action games like pong, snake, or boulder dash.

1.6. Ensuring student success. The most important part of ensuring student success is making sure that each student is working on an individually fruitful and feasible project. It must be fruitful in that the student develops programming and problem solving skills and feels a sense of accomplishment from their work. It must be feasible in that each project must be achievable in the given timeframe with the student's skill set. To that end, the instructor must help each student develop a well-defined project goal of the right scope. This is described in detail in Section 2.3.

During this project work phase, the instructor is simultaneously overseeing 15 to 25 different research projects. For this reason, it is important that the students be able to function mostly independently or with help from their peers. It has been crucial to provide a large amount of scaffolding around every project, including outlining transparent expectations and setting deadlines to keep the students on track. The course webpage provides details about how far along students should be each class period and this is reinforced in class by the instructor. This signals to students who are behind that they need to put in more time to the project and/or come to office hours for extra help.

Because the students have such diverse prior knowledge and backgrounds, the fruitfulness and feasibility of a project will vary from student to student and the grading system must accommodate this disparity. In particular, if project expectations were uniform across all students, this might encourage advanced students to put in minimal work and discourage beginning students. Details about project grading are discussed in depth in Section 3.2.

2. The 3D design project

2.1. Overview. In the second module of the course, students are tasked with designing and 3D printing a mathematical sculpture. Each student artwork is expected to originate from some mathematical concept. The artwork may visualize an idea from mathematics, or instead use mathematical and computational techniques to recreate a specific object or abstract form of interest.

In contrast to more established computer-aided design programs (such as Solidworks, AutoCAD, or Fusion 360), 3D design in Mathematica relies on the user building their model from base mathematical concepts. To be able to do this, students must refresh their mathematical knowledge of three-dimensional coordinate systems, curves, surfaces, and geometric objects.

The author has created Mathematica tutorials on 2D graphics objects, 3D graphics objects, parametric curves and surfaces, transformations thereof, and how to make each of these types of objects 3D printable. (See Appendix A for more information.) These ideas are interwoven with discussions of the elements of art, design thinking, and the 3D printing process.

2.2. Creative expression and artistic intentionality. The author cares deeply that students get a chance to use mathematics creatively and learn to think artistically. A few days into this second module, students are prompted to explore examples of mathematical art online, including works by Bathsheba Grossman[2], Henry Segerman[3], Laura Taalman[4], and various artists exhibiting at the Bridges Mathematical Art Conference[5]. They are asked to choose a few inspiring artworks, and share and annotate them with a few sentences about why they chose the pieces on the online discussion board before class. Here are a few examples of past student comments:

> "I find the following pieces of artwork interesting and inspiring because I can imagine the pieces being utilized as jewelry as a statement piece. The 3D prints in my option are very unique and striking visually."
>
> "This gif unlocked memories I forgot I had. I remembered learning about flowers having symmetry, patterns, and some even follow the Fibonacci sequence. They are mathematical and stunning!"
>
> "[This sculpture] shows the beauty of the randomness and reorganization of water. This piece is not created using a 3D printer, however, I can see something like this being designed and created on a 3D printer."

The next class session the students are in for a treat. A colleague from the art department comes to the classroom and gives a presentation about 3D printing and the artistic elements of sculpture, including the concepts of positive and negative space, symmetry vs. asymmetry, proportion, repetition, contrast, harmony, and movement. After the structured presentation, the class discusses the artistic elements that are present in the mathematical art the students had shared the night before. Because it is often the first time many students have thought about these ideas, they are always engaged and inspired by this discussion.

FIGURE 4. Professor of Art Matthew Greco.

This is supplemented by handouts on the Elements of Art[6] and the Principles of Design[7] from the Getty Museum's education initiative. Once students are given these insights to the world of art and design, they can use this lexicon to express their goals in the creative process. Do they want to aim for a piece that has distinctive negative space? Are they looking for a piece that looks lightweight or instead has gravitas? Through this exposure, students gain the ability to express key aspects of inspiring artwork, become more intentional about how their sculpture develops, and enrich their final reflections.

---

[2]http://www.shapeways.com/shops/bathsheba
[3]http://www.shapeways.com/shops/henryseg
[4]https://mathgrrl.com/designs/
[5]http://gallery.bridgesmathart.org/exhibitions/
[6]https://www.getty.edu/education/teachers/building_lessons/formal_analysis.html
[7]https://www.getty.edu/education/teachers/building_lessons/formal_analysis2.html

2.3. Design thinking. "3D printing" sounds like it could be simple... until you try it. The process involves many messy and complex steps between the initial seed of an idea and the final physical piece. Students confront the full 3D printing experience in this class, including working through productive failure at many levels.

Students are introduced to a macro vision of this adventure from start to finish using the "double diamond" framework in Figure 5. This breaks down the process into four stages: Discover, Define, Develop, and Deliver. The first diamond is where a vague initial concept is made into a concrete problem specification. The second diamond is where the idealized problem specification meets practicality and feasibility to be physically realized.

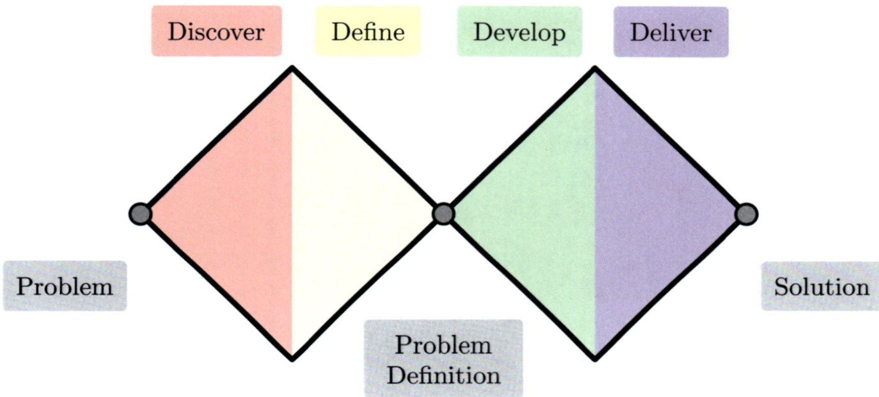

FIGURE 5. The double diamond design model

2.3.1. *Discover stage.* The process starts with the students receiving the very vague project description given in the first paragraph of Section 2.1. The students are primed to think artistically after the web browsing and discussion of the elements of art. At that point, students are encouraged to explore, research, and brainstorm to create many possible project directions that touch on mathematical principles or outside interests.

2.3.2. *Define stage.* This brings us to the widest part of the first diamond and the next stage in the process: problem definition. These many directions created from the discover stage need to be triaged based on personal preference, time constraints, ambition, and peer feedback.

Students propose initial project ideas over email by a specified deadline; the next class period is spent helping students hone these initial ideas to precise project goals. This requires that the instructor understands Mathematica's capabilities and can envision how the student could complete the project when they propose it. Furthermore, the instructor must informally assess each individual student's level of knowledge to make sure that the steps to completion are not too simple nor too complex.

After multiple rounds of back and forth, the result is a precise problem specification that the student can always refer back to if they need to remind themselves what they are trying to accomplish.

2.3.3. *Develop stage.* We have reached the point in between the two diamonds and it is time for students to branch out once again. They need to take the precise idea and understand how they are going to realize it. Students are encouraged to sketch their idea on paper first and develop a plan before starting to code. They then develop the necessary coding skills and mathematical knowledge to explore the space of possibilities. They also make use of a prototyping feedback loop to learn what works and doesn't work in the 3D printing world.

This process has been enhanced by the new Queens College Makerspace, first incorporated into this course in the Fall 2021 semester. Students use the 3D printers for "rapid prototyping" of their initial designs, which can be printed and critiqued within a day's time.

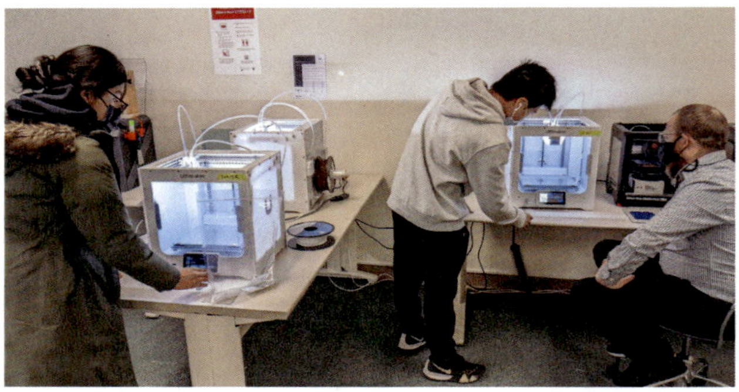

FIGURE 6. Students working in the Queens College Makerspace.

2.3.4. *Deliver stage.* The exploration, research, and prototyping that happens in the develop stage then needs to be harnessed into a final project. As the tests continue, the students give each other feedback until the finalized model is complete. A peer community is extremely beneficial during this process for giving advice about the Mathematica commands they are using and to help shape the final piece. The class period before the project is due is spent doing a formal peer review. The instructor provides pairs of students with a collection of questions that each student answers about the other student's work. The questions guide students to provide constructive feedback about whether the current model, Mathematica notebook, and report satisfy the project expectations.

FIGURE 7. In class peer feedback session.

After instructor feedback on the grading criteria (see Section 3.2), students revise their models one more time before sending the final, honed models to be 3D printed through Shapeways to take advantage of the variety of materials in which they can print, including high resolution nylon and resin, steel, cast metals, and full color sandstone.

2.4. The culminating experience. At the end of the semester, the instructor organizes an art exhibition for students to display their artwork in the Queens College Library. This culminating experience of the semester gives the students an opportunity to take a victory lap and have yet another experience from the art world. The students bring their final 3D printed sculptures and we gather together for every student to give a two-minute artist talk about their piece and the highs and lows of the creation process. It is clear that throughout the process, students have developed agency over their artwork and are proud of the results.

FIGURE 8. A student presenting work during the art exhibition.

FIGURE 9. Students proudly showing off their final artwork, 3D printed in full color sandstone, white nylon, and gold steel.

At the end of the art show, we install the artwork in an art exhibit that stays on display in the library until the end of the next semester. Before the show the instructor requests from the students the name of their artwork and provide a two-sentence description. This information is used to create professional-looking museum placards that are placed in the display next to the artwork. It is empowering to see one's work in an art exhibit alongside other people's creations. Students are always impressed that they have their work in an art exhibit and enjoy being able to show off their artwork to their friends.

The helpfulness of the Queens College Library staff has made this process a pleasure. It is wonderful to be able to hold the event in the library and for the artwork to be displayed in a public space. The exhibits have become a highlight of campus tours for prospective Queens College students.

2.5. Project timeline. When creating a schedule for this module, it is important to account for the time that it takes to prepare a 3D model for printing, have it printed by the 3D printing company, and receive the model after shipping. This explains why this module is the second of three, and not the final one. Here is a representative timeline for organizing the content in a course that meets twice a week for 14 weeks:

FIGURE 10. An installation of student art at the library.

Day 9: The class wraps up Module 1. Students complete a tutorial on the basics of 2D graphics. Students sign up for a tour of the Queens College Makerspace and create a Shapeways for Education account.

Day 10: Students complete a tutorial on the basics of 3D graphics with an emphasis on functional programming. Students are given a list of sites with mathematical art and asked to share and annotate a couple examples on our online discussion board.

Day 11: Students complete a tutorial on more advanced 3D graphics. On this day Professor Greco talks about 3D printing and the artistic elements of sculpture.

Day 12: Students complete a tutorial on making high quality 3D models and start to brainstorm possible project ideas. We discuss how 3D printing works to ensure the final models will be printable.

Day 13: Students complete a tutorial on 3D design techniques. Students finalize their project specifications.

Day 14: Students see a number of minimal working examples to see how to realize common 3D design tasks and watch a video about advanced techniques in 3D design. Students are actively designing their sculpture inside and outside class.

Day 15: Students are in full project work mode inside and outside class. In class we introduce Ultimaker Cura and rapid prototyping, getting practice using the Makerspace to print a model.

Day 16: Students finalize their model and print a physical prototype for peer review. They complete an initial draft of their lab report (minus discussion of revision process).

Day 17: Students complete an in-class peer review of prototype and draft report, after which they revise their sculpture, notebook, and report.

Day 18: Students submit final prototype to Shapeways for 3D printing in desired material, and submit their final report to the instructor. The class transitions to thinking about Module 3.

Soon after Day 18, the instructor gives feedback on the final report and allows for revisions. Shapeways prints and ships the 3D printed model within the next few weeks. The semester culmination event and art exhibit installation take place during our assigned final exam period.

3. Project deliverables and grading

At the end of the project, students submit their physical mathematical sculpture, an organized and commented Mathematica notebook that recreates their 3D model, the digital model computer file, and a two-to-three page writeup about the project.

3.1. The written lab report. Students are required to write and submit a two-to-three page lab report that is a reflection on the artistic and creative process and documents the student's process throughout the project.

The goal is to give students the opportunity to reflect on how they have developed in their roles as programmers, mathematicians, and artists. The report also serves to ensure that the final artwork is a product of deliberate choices and not simple happenstance or academic dishonesty.

From a programming standpoint, students are asked to share the commands and algorithmic techniques they integrate into the coding of their projects and how they advance their programming skills. They also detail how they took the initiative to learn the necessary commands and techniques in Mathematica. Students also need to describe the mathematical principles that are at work, and how those principles are evident in the finished sculpture. Further students address how the mathematical principles match their level of mathematical knowledge.

Students are called upon to reflect upon their artistic choices: the inspiration for the project, the aesthetic they were going for, the elements of art they worked to achieve, the reason for their choice of final material, and what the viewer should appreciate about the artwork.

Students are asked to address prompts related to the creation, prototyping, and revision processes: what sort of obstacles they encountered along the way, how they experienced the first prototype they created and how that inspired them to revise the project. Students have to explain how their project evolved over time, including how discussions with their classmates and during the peer review process influenced the ways in which they changed their project. Last, students are to reflect on how they might approach the process differently if they were starting over.

3.2. Grading the projects. In the author's experience, instructors who exclusively grade mathematics exercises take comfort in the idea that their grading role is objective—a mathematical answer is either right or wrong. (The author disagrees with this assessment.) This can lead to discomfort about one's role in grading student artwork because the universe of possible subjects is infinite and the universe of approaches to realizing the final piece is also infinite. In effect, we are called upon to render a value judgment on student work and at the same time we recognize that our interpretation of all parts of the process is subjective. It is also important to recognize that many math students are also often uncomfortable with the freedom (and therefore the responsibility) that comes from an exercise that does not have one "right" answer.

The author has developed a grading method for the projects in this course that is inspired by his experience in standards-based grading, which he now uses when teaching a non-project-based class like calculus. In standards-based grading, the learning objectives are transparently provided to the students, and regular assessments and reassessments allow students to show their mastery of the material at any point during the semester. Such an approach makes it possible to assess *everything* that the instructor values, including so-called "soft skills" like the ability to discuss math with peers. This method of grading has been transformative because it changes the dynamic in the class from adversarial to collaborative between students and instructor, and grades become much more aligned with student progress towards learning objectives.

The rubric developed for the 3D design project has been refined over the years to address the course learning objectives in Section 1.3. Students are assessed on the following standards.

Attentiveness: The student has made active and steady progress and met deadlines.

Name and Description: The name and description of the final artwork are precise and concise.

Intentionality: The vision of the artwork has been honed over time and has a story behind it.

Mathiness: The object is created using mathematical content at the student's level of knowledge.

Functional Techniques: The object was developed using Mathematica techniques learned in class.

Notebook Organization: The notebook was well organized into sections and with text cells that provide context for the code.

Artistic Writeup: The report discusses the student's development as an artist and artistic choices.

Technical Writeup: The report discusses the student's technical choices and their development as a both a mathematician and programmer.

Revision Writeup: The report discusses the creation, prototyping, and revision processes.

Writeup style: The report is well written and follows the format requirements.

After the student submits their deliverables, each standard is scored on an E-M-R-N scale (Exemplary, Meets Expectations, Revisions Needed, and Not Assessable) and the instructor provides detailed feedback about each standard. The ten scores are converted to a provisional letter grade as follows.

A Earn a score of M or higher on all standards and a score of E on at least eight standards.

B Earn a score of M or higher on all standards and a score of E on at least four standards.

C Earn a score of M or higher on nine standards and no N scores.

D Earn a score of M or higher on eight standards and at most one N score.

F Have fewer than eight E or M scores.

Note that a student cannot receive a grade higher than C if there is some aspect of their project that needs revision. Students are given the opportunity to improve their provisional grade by revising their project deliverables and having the instructor re-score the standards. Due to time constraints and instructor mental load, students are limited to one resubmission.

Consistent with Talbert's pillars of alternative grading (see [**Tal21**]), this grading scheme was developed to provide complete transparency in the expectations for students and how their work will be graded, and by giving students the opportunity for peer and instructor feedback throughout the process with the ability to revise work without penalty.

Students appreciate the transparency in evaluation expectations. Since the expectations align with the class's learning objectives, a student who addresses the items in the rubric is showing that they learned what the instructor wanted them to learn. The revision process benefits both the student and the instructor. The instructor can convey a frank assessment of the student's work and how it meets or does not meet expectations. The student gets a chance to see where their work does not meet expectations and other aspects that could be improved, which reinforces that project development is a process and there is always room for improvement. The student can decide whether it is worth putting in the time to revise their work to improve their standards scores, and in a direct and transparent way, their final project grade.

4. Course reflection

The students and the author enjoy this project immensely; this section includes reflections from both points of view.

4.1. Student comments. Student voices are collected at the end of the semester through a request to write "a letter to future students", a curated collection of which is shared with incoming students the following semester. Here are some of their words of wisdom.

> "For a lot of students, this class might be one of the first instances that unravels for them what it is exactly that is so beautiful about mathematics, though things like precision are not particularly stressed, your creative freedom will know no bounds in this class. You should absolutely take advantage of this, the depth at which you're allowed to pursue a particular problem is seldom allowed in other mathematics courses, this versatility is a gift."

> "Incoming students can expect to be as creative as they want when doing their projects. The professor allows complete creative freedom and that is what made the project fun to complete. It would have been nice to know before hand that I would need to spend the extra hours outside of the classroom to understand the topics. Overall, I have nothing but positive things to say about this class and I would recommend it to anyone who is creative and who wish to show their creative prowess."

> "The amazing thing is that the professor gives you the tool set to do these projects throughout the course and all you have to do is apply that and he even gives you the freedom to put your own flavor in to the mix. With this you get to truly see the diversity that everyone can come up with through each project."

> "We completed some major projects. The most exciting of them all was designing three dimensional objects to be sent to a 3D-printing company

(Shapeways). I was able to make a piece of jewelry based on the intersection of sine waves and lines that I can sell if I wanted to! There's nothing like holding in your hands and feeling the math you've been drawing on paper or seeing on the computer for your entire life. We've built tutorials. We've built games. We've built economic models. We've done so much with the robust power of Mathematica. This class inspired me so much that I want to keep playing with it and exploring it even into the Summer, long after the course is over."

"Do all of the assignments. Play with Mathematica in your spare time. Explore the internet—there are tons of resources, tutorials and communities of users who are already asking questions you probably are searching for the answer to. This is one of the few classes where you'll be challenged to think mathematically, critically, and creatively. This is one of the few classes where you may not know the answer and neither does the professor. This is one of the few classes where you'll learn that it's okay not to have the answer, even in a mathematics class. The professor will work hardest to help you master Mathematica and bring your dreams to live via coding. Hopefully Mathematica will take you to a new place in your journey with mathematics."

"Out of everything I've learned, there's one thing that stuck out to me the most. One day I asked the professor a mathematical question which for me was hard, so I asked him for help. He thought about my question and then told me 'I don't know.' As bright and intelligent as the professor is, he humanized himself in that moment. It made me feel like I have a shot at this math thing. It was more inspirational than anything. He could have given me an answer to make himself look great and leave me baffled but he didn't."

4.2. Instructor reflection. It is always a treat to teach this class. While the project descriptions and expectations have converged over time, the course is always different and exciting to teach! Project deliverables vary greatly from student to student and semester to semester because I encourage students to draw from their lived experience and personal interests and because Mathematica's computational power extends to a vast variety of subject areas.

This course is a lot of work to coordinate and supervise. The students are given leeway to explore, experience productive failure, and take ownership of their learning. This is a messy process and often the first time the students encounter this type of learning environment, so as an instructor I need to encourage students and provide technical help if their project choices end up being a bit too ambitious. This means that in its current incarnation this course cannot realistically scale up to serve more students without additional teaching assistants or peer mentors. Relatedly, because I prioritize ensuring that the day-to-day process runs smoothly, I end up falling behind in returning project feedback promptly.

One conscious choice I made in the course construction is that there are three projects in the semester. I think the first project (the Mathematica tutorial) serves its purpose phenomenally—students learn the structure of Mathematica and its basic data types, they explore a topic of interest, they learn how to format a notebook, and they ease into the self-directed and collaborative modes that make the rest of the semester run smoothly. It also introduces the alternative grading scheme and reinforces that learning and personal growth is valued. Students often wish that they had more time for each project, which is impossible if there are three projects. I take the view that the second project is there to give students a taste of what 3D printing can be. If I devoted a whole semester to it, I am sure that the students' skills in 3D printing would improve as well as the quality of the final pieces. However, we would run into issues at the end of the semester with the time it takes for models to be printed and be shipped, and the students would miss out on the distinct yet complementary third project where they develop different programming skills and learn about designing interfaces with human interaction. Often the second and third projects are both rated equally positively by students in terms of preference and how much they felt they learned.

One of the most rewarding points in the class is at the end of the second project. It is such an amazing and empowering experience to hold something in your hand that had only previously existed in your mind and in the computer. When the final models arrive, students can't wait to show me and their classmates; they regularly relate how proud they were to share their sculpture with their friends and family. And I often get feedback that before the class they never expected to be able to create something of their own, yet after the class feel comfortable and confident with that process. With the opening of the Queens College Makerspace, I expect future students to take even more ownership of their learning.

Just as I hope that my students develop their growth mindset in this course, I am continually working to improve my teaching and the curriculum. For example, each semester I modify the standards and grade conversion table as I learn more about the way I assign scores and what I expect from students. In addition, one thing I noticed that had been lacking in previous semesters was that students would proceed linearly from start to finish without exploring a wide space of possible solutions. So I added a requirement in Fall 2021 that students need to modify some parameter or aspect of their project, and reflect on which version they prefer and why. With this prompt, students thought more deeply about why they ended up with their final result and they shared more about their creative process in their reports.

Mathematical Computing is a course destined for math majors who have a certain amount of mathematical background. I wanted to be able to offer the opportunity to be creative with mathematics and create mathematical art to non-majors. I created a project-based course for non-majors called Mathematical Design with the online graphing calculator Desmos as its medium. I look forward to further developing the curriculum for this class and publishing more about it. Stay tuned!

Appendix A. 3D design in Mathematica

In this section the reader can learn multiple techniques to use Mathematica for 3D design. An interactive copy of this notebook is available on the Wolfram Notebook Archive [**Han22**]. Additional resources about 3D design in Mathematica are available on the author's webpage[8].

The author would like to acknowledge Mathematica's Documentation Center [**Wol22**] for its extensive examples that explain the Wolfram Language in great detail, the Mathematica StackExchange community[9] for their selfless sharing of ideas that have inspired a number of advanced explorations, and Henry Segerman's article [**Seg12**], which was very helpful with initial trials and tribulations in 3D modeling in Mathematica. Lastly, many thanks to Jakub Kuczmarski for his `mmaCells` LaTeX package[10] that allowed for in-line typesetting of Mathematica code in this article.

A.1. Basic 3D graphics and exporting. Creating a 3D model file with Mathematica can be exceptionally easy or extremely frustrating. In many cases, creating your 3D file is but a simple call to the `Export` function, where you specify the file extension `.stl` for uncolored models or `.wrl` for colored models.

```
In[1]:= Export["file.stl", model]\index{export}
```

In general, I like to append the file name to the current notebook's working directory (using the `NotebookDirectory` command) to organize my files.

The most basic way to create an object in Mathematica is to specify the coordinates and parameters of the 3D primitives such as `Sphere`, `Cylinder`, `Tube`, `Tetrahedron` and wrap them all in a `Graphics3D` command. Happily, starting in Mathematica version 13, these primitives behave well when they are exported. For example, here is the code that builds a model of a basic snowman using spheres, cylinders, a cone, and some rectangular prisms:

```
In[2]:= snowman = Graphics3D[{
          Sphere[{0, 0, 0}, 1],
          Sphere[{0, 0, 1.3}, 0.8],
          Sphere[{0, 0, 2.4}, 0.6],
          Black, Cylinder[{{0, 0, 2.8}, {0, 0, 2.9}}, 0.7],
          Cylinder[{{0, 0, 2.9}, {0, 0, 3.5}}, 0.5],
          Orange, Cone[{{0, -0.55, 2.4}, {0, -0.9, 2.4}}, 0.1],
          Black, Cuboid[{0, -0.75, 1.3} - 0.1 {1, 1, 1},
                        {0, -0.75, 1.3} + 0.1 {1, 1, 1}],
          Cuboid[{0, -0.7, 1.6} - 0.1 {1, 1, 1},
                 {0, -0.7, 1.6} + 0.1 {1, 1, 1}],
          Cuboid[{0, -0.7, 1.0} - 0.1 {1, 1, 1},
                 {0, -0.7, 1.0} + 0.1 {1, 1, 1}]
        }]
        Export[NotebookDirectory[] <> "snowman.stl", snowman]
```

In each `Sphere` command, you specify the center of the sphere and its radius. In the `Cylinder` and `Cone` commands you are specifying the endpoints of its axis of

[8] `http://qc.edu/~chanusa/mathematica`
[9] `https://mathematica.stackexchange.com/`
[10] `https://github.com/jkuczm/mmacells`

rotation as well as its radius. In the `Cuboid` command, you specify the opposite corners of the rectangular prisms. The resulting snowman looks like this:

Out[2]=

I always suggest that you re-import your STL files to ensure that everything exported correctly.

```
In[3]:= Import[NotebookDirectory[] <> "snowman.stl"]
```

Out[3]=

Upon observation, we see that the imported model has no color and that the final result is very faceted because the model was exported with the default options. We can greatly improve the result with the more advanced commands shown in Section A.4; we use these techniques to write updated code to create a high quality, full color model of the snowman in Section A.8.

Mathematica is not currently capable of importing WRL files so any colored models should be opened in a different program or uploaded to a web service such as Sketchfab[11] to ensure that the model has exported correctly.

A.2. Curves. The next level of 3D design complexity uses knowledge of multivariable calculus to create models of curves and surfaces. A 3D model of a vector curve $\mathbf{r}(t) = \langle f(t), g(t), h(t) \rangle$ is made using `ParametricPlot3D` with one input variable. For example, a basic helix can be created using the code

```
In[4]:= ParametricPlot3D[{t/Pi, Cos[t], Sin[t]}, {t, 0, 6 Pi}]
```

Out[4]=

However, this is where the precision of an idealized mathematical model runs into the reality of what can be 3D printed. A helix is a one-dimensional curve embedded in three dimensions, so we need to add some thickness to create a printable model. We can do this using a `PlotStyle` option that specifies the radius of a tube that will be swept out along the curve.

```
In[5]:= helix = ParametricPlot3D[{t/Pi, Cos[t], Sin[t]},
          {t, 0, 6 Pi}, PlotStyle -> {Tube[0.1]},
          Axes -> False, Boxed -> False, SphericalRegion -> True];
        Import[Export[NotebookDirectory[] <> "helix.stl", helix]]
```

Out[5]=

The `Axes`, `Boxed`, and `SphericalRegion` options improve visualization in the notebook by removing the default framing around the model and ensuring that when you rotate the model it does not resize the cell.

It is only on re-importing the model do we realize that the result is once again very faceted. We can use the `PlotPoints` option twice to improve the quality of this 3D print. We specify how many points will be sampled along the curve (here 500), as well as how many points will be sampled in the circular cross-section of the tube (here 50).

```
In[6]:= helixHD = ParametricPlot3D[{t/Pi, Cos[t], Sin[t]},
          {t, 0, 6 Pi}, PlotStyle -> {Tube[0.1, PlotPoints -> 50]},
          PlotPoints -> 500,
          Axes -> False, Boxed -> False, SphericalRegion -> True];
        Import[Export[
          NotebookDirectory[] <> "helix.HD.stl", helixHD]]
```

[11]`https://sketchfab.com`

Out[6]=

(In your notebook the first `PlotPoints` option will be red.) The more points specified, the more detailed the print will be; at the same time, this increases computation time and the size of the exported file. The two plots `helix` and `helixHD` look the same in the notebook; it is only upon re-importing the stl models that we can see that the fidelity of the model has greatly improved.

`ParametricPlot3D` also allows you to 3D print a spline specified by a set of control points. The primitive for a spline is a `BSplineCurve`:

```
In[7]:= pts = {{0, 0, 0}, {1, 0, 0}, {0, 1, 0}, {0, 0, 1}};
        Graphics3D[{
          Blue, Map[Sphere[#, 0.03] &, pts],
          Thick, Green, Line[Append[pts, pts[[1]]]],
          Red, BSplineCurve[pts, SplineClosed -> True]},
        Axes -> True, AxesOrigin -> {0, 0, 0}, Boxed -> False]
```

Out[7]=

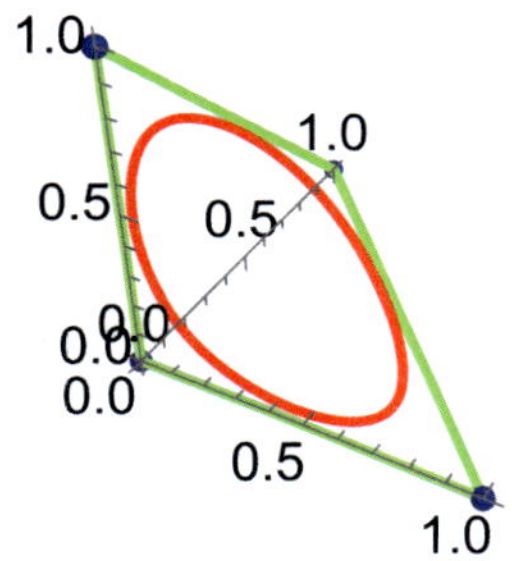

You can 3D print this curve by first converting the primitive to a parametric function on the domain $[0,1]$ by replacing `BSplineCurve` by `BSplineFunction` and then using `ParametricPlot3D` as above.

```
In[8]:= ParametricPlot3D[
          BSplineFunction[pts, SplineClosed -> True][t],
          {t, 0, 1}, PlotStyle -> {Tube[0.02]},
          PlotRange -> All, Axes -> False, Boxed -> False]
```

Out[8]=

A.3. Surfaces. There are four ways to create two-dimensional surface embedded in three dimensions, depending on how it is defined.

When the surface is the graph of a function $f(x, y)$ of two variables, use `Plot3D`. You need to specify the two independent variables and their domains. By default

`Plot3D` only outputs the 2D surface; thicken it to make it 3D printable by specifying the `PlotTheme` option to be `"ThickSurface"`.

```
In[9]:= Plot3D[Sin[x + y^2], {x, -3, 3}, {y, -2, 2}, Mesh -> None,
          PlotPoints -> 30, PlotTheme -> "ThickSurface"]
```

Out[9]=

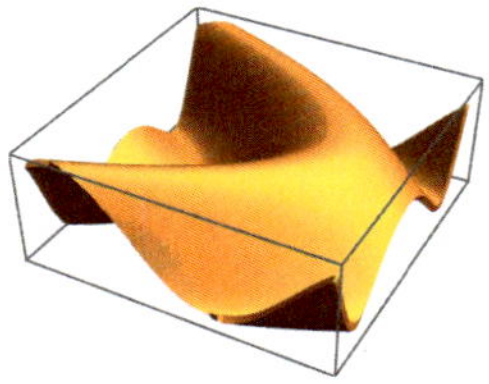

That option does not allow you to specify a thickness; to do so instead use the hidden `Extrusion` option to thicken the surface in the direction normal to the surface. (This option will show in red in your notebook.) In the following example we also see the domain does not have to be a rectangle. (In your Mathematica code be sure to replace IN by ∈ by typing esc \in esc.)

```
In[10]:= Plot3D[Sin[x + y^2], {x, y} IN Disk[{0, 0}, 2],
           Mesh -> None, PlotPoints -> 30, Extrusion -> 0.25]
```

Out[10]=

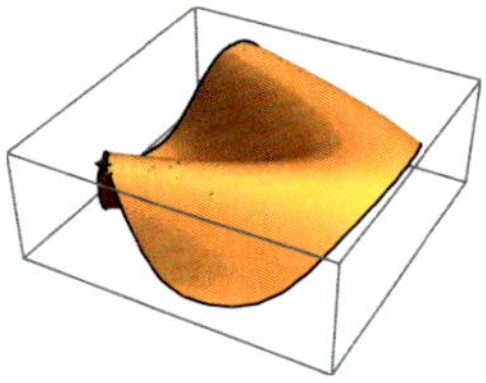

If instead the surface is defined implicitly as the level surface of a function $f(x, y, z) = k$, use `ContourPlot3D`:

```
In[11]:= ContourPlot3D[x^3 + y^2 - z^2 == 0,
           {x, -2, 2}, {y, -2, 2}, {z, -2, 2},
           Mesh -> None, PlotPoints -> 30, Extrusion -> 0.2]
```

Out[11]=

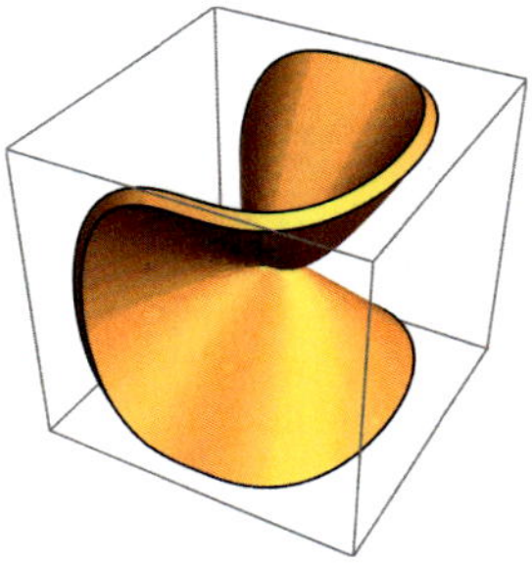

Otherwise, if the surface is defined parametrically as a function $f(u, v)$ of two parameters, use `ParametricPlot3D` and specify the domain of each parameter. The following example is a torus; we specify the number of sample points for each parameter independently.

```
In[12]:= ParametricPlot3D[
           {(3 + Cos[v]) Cos[u], (3 + Cos[v]) Sin[u], Sin[v]},
           {u, 0, 2 Pi}, {v, 0, 2 Pi},
```

```
    Mesh -> None, PlotPoints -> {100, 30}]
```

Out[12]= 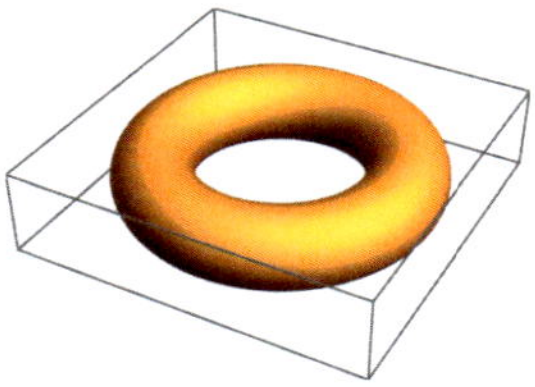

That torus is the boundary of a 3D solid, so the file exported by Mathematica can be 3D printed. In general, a parametric surface may not be the boundary of a solid so the slicing software for 3D printing will tell you that your object is "not manifold". You will need to thicken those surfaces as we did before.

```
In[13]:= ParametricPlot3D[
           {v Cos[u], v Sin[u], -Cos[v]},
           {u, 0, 2 Pi}, {v, 0, 2 Pi},
           Mesh -> None, PlotPoints -> {100, 30}, Extrusion -> .2]
```

Out[13]= 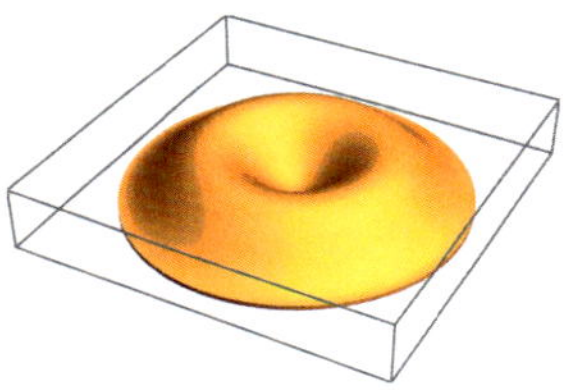

Last, you can create a two dimensional spline surface using `BSplineSurface`. Here is the example from the Wolfram Documentation Center that generates a random surface over the rectangular domain $[1, 5] \times [1, 5]$.

```
In[14]:= pts = Table[{i, j, RandomReal[{-1, 1}]}, {i, 5}, {j, 5}];
         Graphics3D[BSplineSurface[pts]]
```

Out[14]= 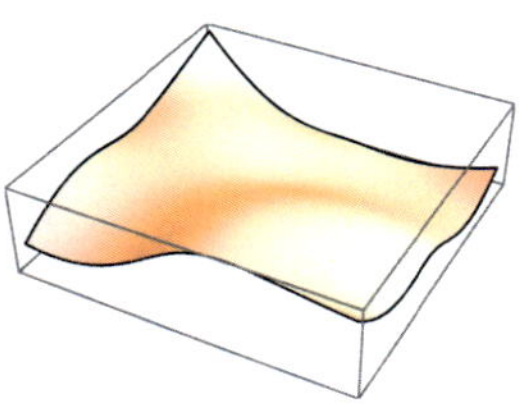

Again we need to convert the `BSplineSurface` to a `BSplineFunction` to create a 3D printable model.

```
In[15]:= ParametricPlot3D[BSplineFunction[pts][u, v],
           {u, 0, 1}, {v, 0, 1}, Mesh -> None, Extrusion -> .2]
```

Out[15]= 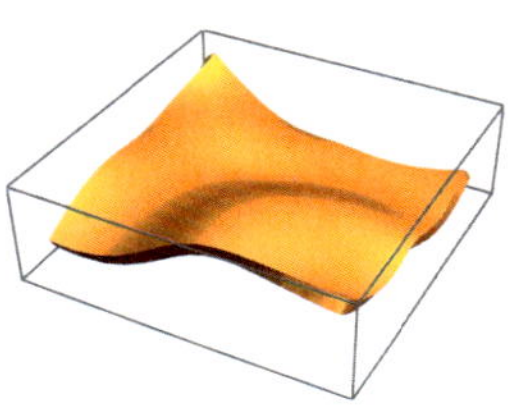

A.4. Getting to yes: meshes and manual discretization. Mathematica uses mesh-based geometric regions to make three-dimensional objects 3D printable. This is similar to how complex numerical expressions are converted to a decimal

```
In[16]:= N[Sqrt[5 + Sqrt[3]]]

Out[16]= 2.59462
```

or how a two-dimensional scene is represented as a collection of pixels.

```
In[17]:= Rasterize[RegularPolygon[7], ImageSize -> 100]
```

Out[17]=

The process of approximating an ideal object by an approximation is called discretization (or triangulation). This replaces the smooth object by a discrete object made up of cells: points, line segments, triangles, tetrahedra, and higher-dimensional simplices. There are two main ways to describe a discretized object: either by the cells that comprise it or by the cells of its boundary. Compare the following two discretizations of the circle:

```
In[18]:= {DiscretizeRegion[Disk[]],
          BoundaryDiscretizeRegion[Disk[]]}
```

Out[18]=

In the first example, the circle is approximated as a collection of two-dimensional triangles. In the second example, the circle is approximated as the interior of a collection of one-dimensional line segments. In the same way, a three-dimensional form can be represented as a collection of three-dimensional tetrahedra or as the interior of a collection of two-dimensional triangles.

When you discretize a graphics object, use `DiscretizeGraphics` or `BoundaryDiscretizeGraphics`. When you discretize a shape that is defined as a region, use `DiscretizeRegion` or `BoundaryDiscretizeRegion`. The resulting objects are `MeshRegion` objects or `BoundaryMeshRegion` objects, respectively.

Mathematica has a large number of built-in polyhedra, accessible using the command `PolyhedronData`; their meshes can be accessed directly from the command by specifying `"MeshRegion"` or `"BoundaryMeshRegion"`.

```
In[19]:= PolyhedronData["Dodecahedron", "BoundaryMeshRegion"]
```

Out[19]=
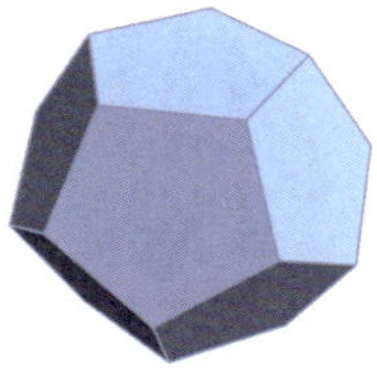

Since boundary representations are one lower dimension than cell representations, they are computationally simpler so more of Mathematica's functionality applies. However, not all shapes have a nice boundary representation; in that case you will have to work with the shape's cell representation. Wolfram developers have been continually improving these commands; great strides have been made in the past few years.

When you use `Export`, (I think) Mathematica tries its best to discretize each piece of your model, then combines everything together. This process sometimes goes awry, so it may make sense to discretize your objects first, use a `Show` command to combine them, and export the result. Do note that it may be beneficial to discretize every single object individually instead of trying to discretize a collection of objects. Compare the following.

```
In[20]:= balls = Table[
            Sphere[{i, j, 0}, RandomReal[]], {i, 10}, {j, 10}];
         {DiscretizeGraphics[balls],
            Show[Map[DiscretizeGraphics, balls]]}
```

Out[20]=
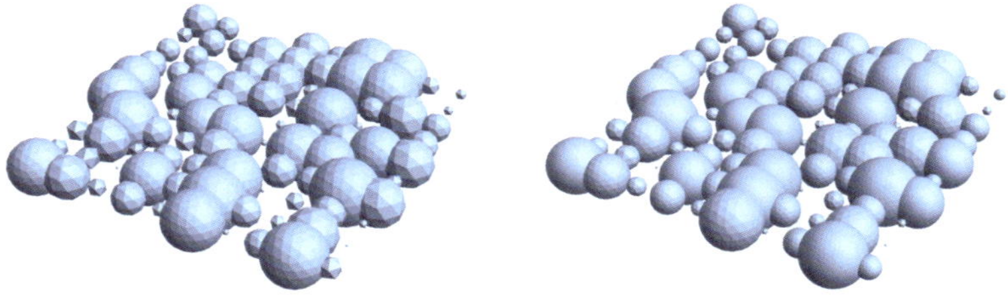

The resolution of the second code block is better because `DiscretizeGraphics` is being applied to each sphere individually.

Furthermore, because `Export` appears to try much harder than the various `Discretize` commands, sometimes it is even beneficial to export and re-import each piece of the model and assemble those pieces together.

It is possible to specify the quality of a discretization of a 3D model by applying the `MaxCellMeasure` option which specifies the maximum size for each highest-dimension cell. Compare:

```
In[21]:= {BoundaryDiscretizeGraphics[Ball[], MaxCellMeasure->.01],
          BoundaryDiscretizeGraphics[Ball[], MaxCellMeasure->.001],
          BoundaryDiscretizeGraphics[Ball[], MaxCellMeasure->.0001]}
```

Out[21]= 

As of Mathematica version 13, the various region objects can be imported into a `Graphics3D` scene as primitives, which makes it possible to finalize each individual piece of the model and then color them independently for export as a WRL file, which is how the snowman is improved in Section A.8.

In my experience, applying some combination of these techniques allow the vast majority of models to be exported by Mathematica. If you are still having issues, see Section A.7 for another tool.

A.5. Extracting mesh information. There are two useful commands to situate your object in 3D space. `RegionBounds` gives the smallest and largest x, y, and z values over the whole object. (Do not confuse `RegionBounds` with `RegionBoundary` which finds a region's geometric boundary.) And `RegionCentroid` gives the centroid of the object. We observe that the following discretized ellipse is 3.8 units wide and is centered at $(0, 0, 0)$.

```
In[22]:= mesh = BoundaryDiscretizeGraphics[
           Ellipsoid[{0, 0, 0}, {2, 1, 1}], MaxCellMeasure -> 0.2];
         RegionBounds[mesh]
         Map[#[[2]] - #[[1]] &, %]
         Chop[RegionCentroid[mesh]]
```

```
Out[22]= {{-1.90211, 1.90211}, {-1., 1.}, {-1., 1.}}
         {3.80423, 2., 2.}
         {0, 0, 0}
```

As we build complexity into our models, it becomes useful to work directly with the internal structure of the mesh. Use `MeshCoordinates` to extract the coordinates of the mesh and `MeshPrimitives` to extract higher-dimensional pieces of the object, like its edges or faces. The second input is the dimension of the parts you want to extract.

```
In[23]:= coords = MeshCoordinates[mesh];
         edges = MeshPrimitives[mesh, 1];
         faces = MeshPrimitives[mesh, 2];
```

These commands allow you to efficiently make a wireframe version of your object or algorithmically color the faces of your object.

```
In[24]:= thickness = 0.06;
         Graphics3D[{
           Map[Sphere[#, thickness] &, coords],
           Map[Tube[#, thickness] &, edges]
```

```
}, Boxed -> False, SphericalRegion -> True]
```

Out[24]=

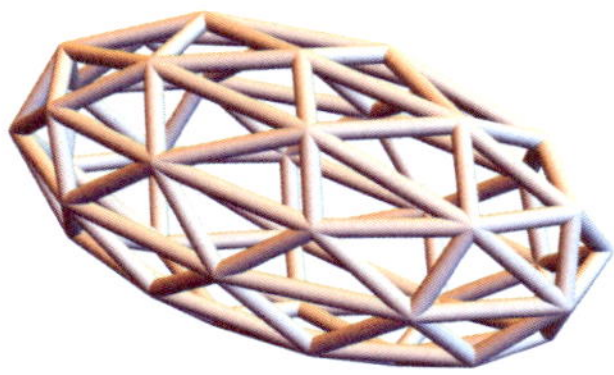

```
In[25]:= colors = Map[Hue[(RegionCentroid[#][[3]]+1)/2] &, faces];
         Graphics3D[MapThread[{#2, #1} &, {faces, colors}],
           Boxed -> False, SphericalRegion -> True]
```

Out[25]=

Note that the wireframe object needs to include spheres at each vertex so that no gaps occur where two cylinders meet. Compare the 3D models generated with spheres and without:

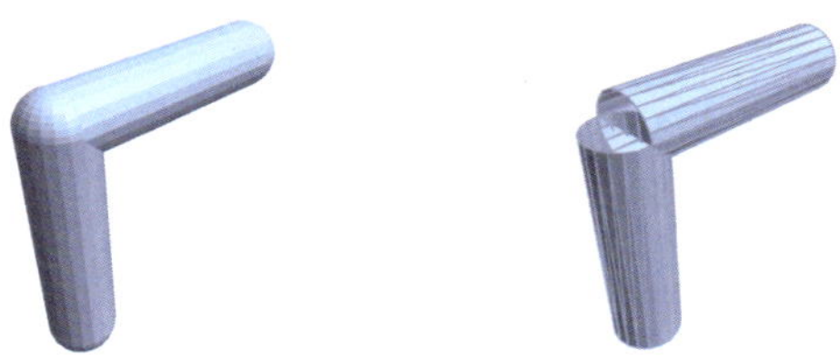

The model without spheres only consists of the shell of each cylindrical edge. Because it is not solid, it would not be 3D printable.

A.6. Region operations. Once you have discretized your models (or pieces of your models) using the discretizing commands or by exporting and reimporting, you have `RegionMesh` or `BoundaryRegionMesh` objects and you can apply various region operations to them.

It is important to be able to change the size and orientation of your model. For instance, it may be easiest to work in one coordinate system to create your mathematical form and in a different coordinate system when printing. In addition, you will want to specify the dimensions of the piece in millimeters if you are uploading your model to a 3D printing service.

To change the size of your model use `RegionResize`. If your second input is one number, Mathematica will scale the object to have first coordinate equal to that input, respecting the existing box ratios. If you instead specify a list of parameters, it will scale the object to fit into the given box, not respecting the existing box ratios.

If we want our model to be 100mm wide, we would write

```
In[26]:= mesh = PolyhedronData["Dodecahedron", "BoundaryMeshRegion"]
         RegionResize[mesh, 100]
```

Out[26]=
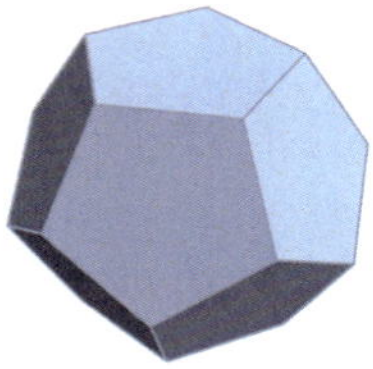

or if we want to skew it to fit in a box that is 80mm × 50mm × 30mm, we would write

```
In[27]:= mesh = PolyhedronData["Dodecahedron", "BoundaryMeshRegion"]
         RegionResize[mesh, {80, 50, 30}]
```

Out[27]=
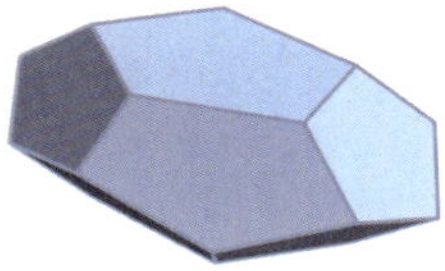

Use `TransformedRegion` and the various `Transform` commands to apply transformations to a region. Suppose we want to apply transformations to the basic cube:

```
In[28]:= cube = BoundaryDiscretizeGraphics[Cuboid[]];
```

To rotate the object by a given angle θ around an axis of rotation (with direction v and passing through point p), input these values into `RotationTransform`.

```
In[29]:= rotatedcube = TransformedRegion[cube,
           RotationTransform[Pi/4, {0, 0, 1}, {1/2, 1/2, 0}]];
         Graphics3D[{Green, cube, Red, rotatedcube},
           Boxed -> False]
```

Out[29]=
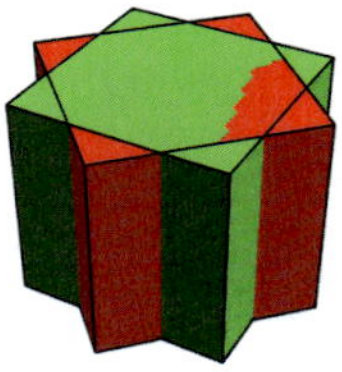

To translate the object along a vector v, input that vector into `TranslationTransform`.

```
In[30]:= translatedcube = TransformedRegion[cube,
           TranslationTransform[{0.5, 0.5, 0.5}]];
         Graphics3D[{Green, cube, Red, translatedcube},
           Boxed -> False]
```

Out[30]=
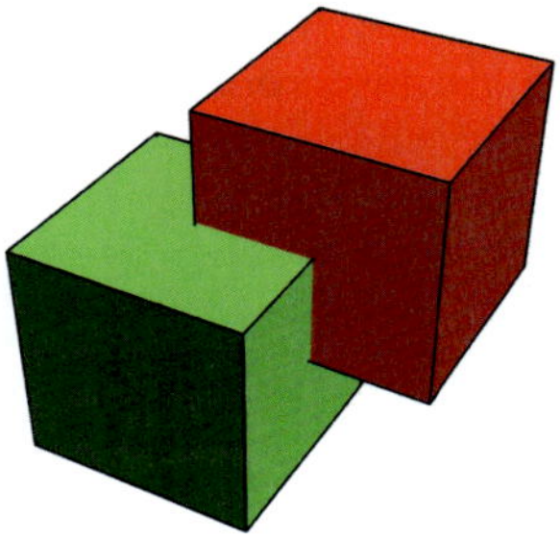

Combining a few of these techniques is useful to orient your object. For example, if you want to move the center of your object to the origin, apply a translation equal to the negative of its centroid:

```
In[31]:= TransformedRegion[mesh,
           TranslationTransform[-RegionCentroid[mesh]]],
```

To intersect, merge, or subtract two regions, use `RegionIntersection`, `RegionUnion`, and `RegionDifference`. These commands work well in 2D. On the other hand, it has been hit or miss whether these commands work on 3D MeshRegion objects. Starting in version 12.1 and further improved in version 13 of Mathematica, these boolean operations work more often for 3D boundary representations.

```
In[32]:= ball = BoundaryDiscretizeGraphics[
           Graphics3D[Ball[{0, 0, 0}, 1]]];
         cube = BoundaryDiscretizeGraphics[
           Graphics3D[Cuboid[{0.1, -0.5, 0.1}, {1, 0.5, 1}]]];
         dodec = RegionResize[
           PolyhedronData["Dodecahedron", "BoundaryMeshRegion"], 2];
         {RegionUnion[ball, cube],
          RegionUnion[dodec, cube],
          RegionIntersection[ball, cube],
          RegionDifference[ball, cube],
          RegionDifference[dodec, cube]}
```

Out[32]= 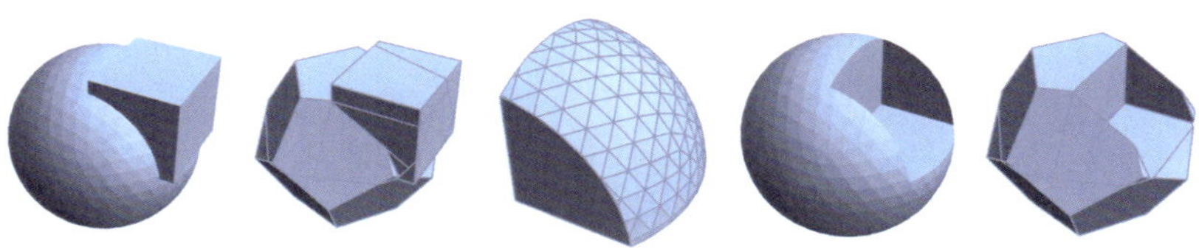

A.7. RegionPlot3D. Another command that can achieve high resolution models is `RegionPlot3D`. It can be very computationally intensive but it is sometimes the only command that works. This command takes as input a boolean expression that defines your desired region and the domain that you want sampled. `RegionPlot3D` gives all the points in the specified domain that satisfy your given requirements.

For example, if you want to find the intersection of two unit spheres centered at $(0, 0, 0)$ and $(1, 0, 0)$, you specify the inequality that defines each sphere and use `&&` (the syntax for 'and') to indicate that you want their intersection, where both inequalities are true.

```
In[33]:= RegionPlot3D[
           (x^2 + y^2 + z^2 <= 1) && ((x - 1)^2 + y^2 + z^2 <= 1),
           {x, -1, 2}, {y, -1.5, 1.5}, {z, -1.5, 1.5}]
```

Out[33]= 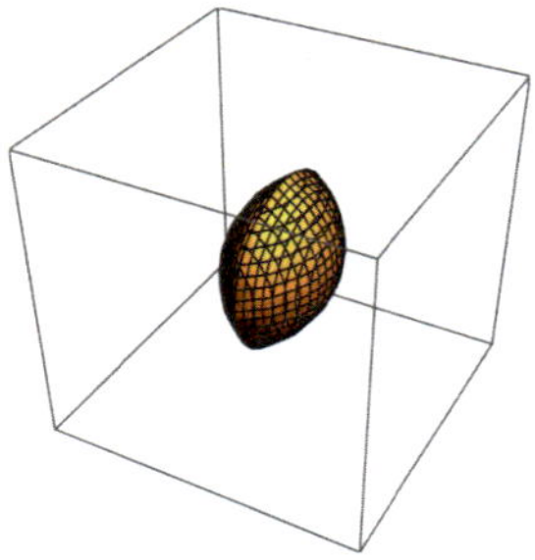

or use `||` (the syntax for 'or') to indicate that you want their union, where at least one of the two inequalities is true.

```
In[34]:= RegionPlot3D[
           (x^2 + y^2 + z^2 <= 1) || ((x - 1)^2 + y^2 + z^2 <= 1),
           {x, -1, 2}, {y, -1.5, 1.5}, {z, -1.5, 1.5}]
```

Out[34]= 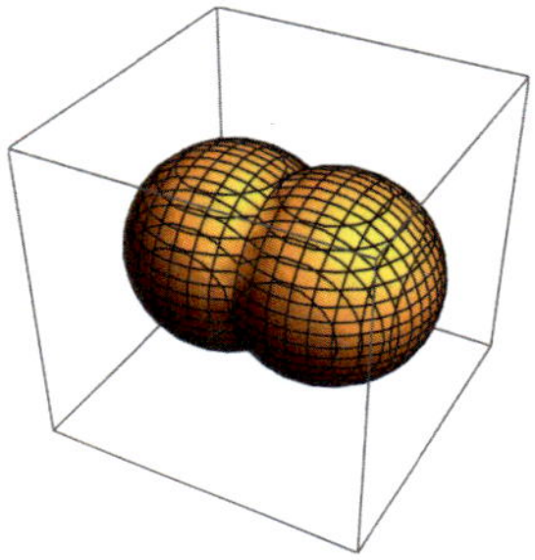

RegionPlot3D is a nice way to create 3D models of surfaces defined by an equation $z = f(x, y)$ that have a flat base for display. We do that by specifying the region as the set of points that lie below the curve.

```
In[35]:= block = RegionPlot3D[Sin[x] + Sin[y] > z,
           {x, 0 - Pi/2, 6 Pi - Pi/2}, {y, 0 - Pi/2, 6 Pi - Pi/2},
           {z, -3, 3}, BoxRatios -> Automatic];
         Import[Export[NotebookDirectory[] <> "Bumpy.stl", block]]
```

Out[35]= 

I assign the option BoxRatios to Automatic so that the axes are scaled to ensure the unit distance in each direction is the same.

We see that the quality of the surface could be improved. We can specify the number of points that Mathematica should sample in each direction using the PlotPoints option.

```
In[36]:= blockHD = RegionPlot3D[Sin[x] + Sin[y] > z,
           {x, 0 - Pi/2, 6 Pi - Pi/2}, {y, 0 - Pi/2, 6 Pi - Pi/2},
           {z, -3, 3}, PlotPoints -> {100, 100, 100},
           BoxRatios -> Automatic];
```

```
Import[Export[
  NotebookDirectory[] <> "Bumpy.HD.stl", blockHD]]
```

Out[36]=

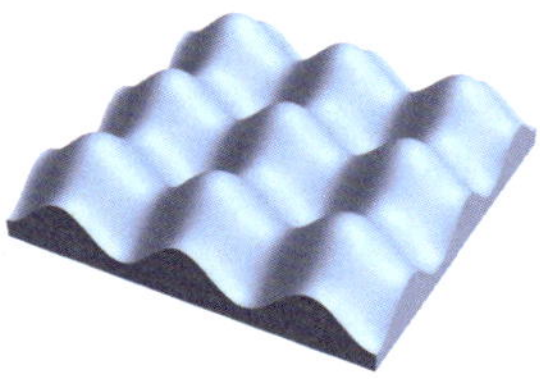

The more points you specify, the better the result. However, this becomes computationally intensive very quickly as you increase the number of points and it rarely produces as precise of a model as we had using the earlier methods.

A.8. High quality snowman. Below is code that creates a high quality, full color 3D model of the snowman from Section A.1. It took multiple iterations to make it work—I ran into errors to export to a `.wrl` file until I converted every mesh to its polygonal representation, hence the use of `MeshPrimitives`.

```
In[37]:= snowball1 = MeshPrimitives[BoundaryDiscretizeRegion[
           Ball[{0, 0, 0}, 1], MaxCellMeasure -> .0005], 2];
         snowball2 = MeshPrimitives[BoundaryDiscretizeRegion[
           Ball[{0, 0, 1.3}, 0.8], MaxCellMeasure -> .0005], 2];
         snowball3 = MeshPrimitives[BoundaryDiscretizeRegion[
           Ball[{0, 0, 2.4}, 0.6], MaxCellMeasure -> .0005], 2];
         tophat1 = MeshPrimitives[BoundaryDiscretizeRegion[
           Cylinder[{{0, 0, 2.8}, {0, 0, 2.9}}, 0.7],
           MaxCellMeasure -> 0.0005], 2];
         tophat2 = MeshPrimitives[BoundaryDiscretizeRegion[
           Cylinder[{{0, 0, 2.9}, {0, 0, 3.5}}, 0.5],
           MaxCellMeasure -> 0.0005], 2];
         carrot = MeshPrimitives[BoundaryDiscretizeRegion[
           Cone[{{0, -0.55, 2.4}, {0, -0.9, 2.4}}, 0.1],
           MaxCellMeasure -> .00005], 2];
         button1 = MeshPrimitives[BoundaryDiscretizeGraphics[
           Cuboid[{0, -0.75, 1.3} - 0.1 {1, 1, 1},
                  {0, -0.75, 1.3} + 0.1 {1, 1, 1}]], 2];
         button2 = MeshPrimitives[BoundaryDiscretizeGraphics[
           Cuboid[{0, -0.7, 1.6} - 0.1 {1, 1, 1},
                  {0, -0.7, 1.6} + 0.1 {1, 1, 1}]], 2];
         button3 = MeshPrimitives[BoundaryDiscretizeGraphics[
           Cuboid[{0, -0.7, 1.0} - 0.1 {1, 1, 1},
                  {0, -0.7, 1.0} + 0.1 {1, 1, 1}]], 2];
         snowmanHD = Graphics3D[{EdgeForm[None],
           White, snowball1, snowball2, snowball3,
           Black, tophat1, tophat2, button1, button2, button3,
           Orange, carrot},
           Boxed -> False, SphericalRegion -> True]
         Export[NotebookDirectory[] <> "snowmanHD.wrl", snowmanHD]
```

You can interact with the final model on Sketchfab at `https://skfb.ly/ovIVI`.

Acknowledgments

Many thanks to Leah Wrenn Berman, Emily Dennett, and Patrick Johnson for their advice for improving this article. I would also like to thank Matthew Greco for being a great collaborative partner at Queens College.

References

[Han22] Christopher Hanusa, "3D Design in Mathematica" from the Notebook Archive (2022), `https://notebookarchive.org/2022-06-dtubqzg`

[Seg12] H. Segerman, *3D printing for mathematical visualisation*, Math Intelligencer **34**, (2012) 56–62. `https://doi.org/10.1007/s00283-012-9319-7`

[Tal21] R. Talbert, *Finding common ground with grading systems*, Grading for Growth, `https://gradingforgrowth.com/p/finding-common-ground-with-grading`, August 30, 2021.

[Wol22] Wolfram Research, Inc., Mathematica, Version 13.0.1, Champaign, IL (2022).

Department of Mathematics, Queens College (CUNY), 65-30 Kissena Blvd., Flushing, New York 11367

Email address: `chanusa@qc.cuny.edu`

Proceedings of Symposia in Applied Mathematics
Volume **79**, 2023
https://doi.org/10.1090/psapm/079/00716

Teaching 3D printing and mathematics at a small public liberal arts college

Ivan Sterling

ABSTRACT. This article is about *3D Printing and Mathematics* courses and independent projects I have taught or mentored since Fall 2016. My goal is to convey the remarkable overall success of these classes and projects. The students displayed excitement about their projects, showed imagination and originality, and learned a wide variety of skills. In this article we've selected samples which include soap bubbles, geometric fabrics, an ellipsoidal dancer, packing tetrahedra into a sphere, and a calculus toolbox. For readers who are considering teaching a 3D Printing course, my hope is that this article will be of some help.

1. Overview

The topics course *3D Printing and Mathematics* was offered in the Falls of 2016, 2018, and 2021. It was listed as a four-credit 400 level mathematics class and could be used as an elective toward fulfilling the mathematics major requirements. The prerequisite was Vector Calculus and nearly all the students were mathematics majors. Each course had between ten and twelve students. In addition about a dozen students continued on to a four credit Senior Projects involving 3D printing. This article focuses on five or six projects, out of some thirty or forty that students have completed. All the students were undergraduates.

I am mathematician in the Mathematics and Computer Science Department at St. Mary's College of Maryland (a non-religious public four years honors college). My specialty is differential geometry and at one time I did extensive work in surface theory computer visualization. 3D printing is one of many ways to fabricate mathematical objects. My first experience with mathematical fabrication came around 1988 when I played a minor role in the creation of a large metal sculpture of a special Boy's surface. The sculpture, Figure 1, permanently resides at Oberwolfach, a Mathematics Research Institute (MFO) in Germany. On January 28th, 1991, it was installed at the MFO as a gift of Mercedes-Benz.

I was frustrated by the poor resolution inherent to low end 3D printers and tried for several years to secure funding for a "real" 3D printer such as the basic Ultimaker. Finally, in a joint venture with the Physics Department we were able to purchase one Ultimaker printer. It was with this printer that I ran my first 3D printing course in the Fall of 2016. I've always been interested in ways that

Photograph Credits: Art Students Sydney Rourke and Tony Rayo.

FIGURE 1. Willmore's boy surface at Oberwolfach

3D printing can be used to aid in mathematical research and used to supplement teaching mathematics at nearly all levels. Some of the prints mentioned below are displayed at the Mathematics table on admissions days and open house days.

An interesting side story on the utility of 3D printed models in learning: A student in my complex analysis class wanted to 3D print some of the functions we had been learning. They started with exp(z) and then they did log(z). They were surprised that when they held the two prints in their hands there was no way to tell them apart. After a minute, I pointed out that the same thing happens in pre-Calculus. As 3D prints of exp(x) and log(x) are identical. This is no surprise as they differ by a simple rigid motion in the plane; reflection across the diagonal. Not so obvious is that the same is true for the graphs of exp(z) and log(z) as surfaces in 4-space. Having easy access to a 3D printer makes such side stories possible.

2. Course structure and considerations

The classes met twice a week for about two hours. The students in these courses were there by choice and most were so devoted to their projects that they committed significant time outside of class to ensure success. When progress was slow, we would typically try an easier version of the project or attempt the same project with a wider variety of settings.

Nearly every class started with each student (or team) giving a brief summary of what challenges they were facing, how they tried to address the challenges, what worked and what didn't. Over the years the basic structure of the course didn't change. In the early courses, we spent more time on surveying the various types of software available, and in the later courses, more time was spent on programming in Mathematica. In the first course there was only one printer, for the second there were two printers. We started the Fall 2021 course with three printers and ended with five. There was only one main project for the Fall 2021: to build a discrete hyperbolic surface prototype (see definition below).

The goal of the first day of class was to show students how to print an object. By the second class they were expected to have printed one. Typically these first prints only required the students to find an STL file on the internet. Slice it in a the machine's software, typically CURA, which creates GCODE. Save the GCODE on a flash drive, put the flash drive into the printer, and hit print. Naturally this is only the beginning print and for projects they needed to master many important methods and settings. Once students began working on their own projects challenges always arose. Students overcame obstacles by searching the documentation, discussing and debating with fellow students, and asking for my help.

Students came to the courses with a wide variety of experiences in 3D printing and in mathematics. For example in one course I had a student with a 3D printer at home who had been printing on a regular bases for years. Another student was a transfer student with an architectural background. Some students prefer "working with their hands", etc. I'll assume this is also true of the audience for this article. I tried to adjust my pedagogical techniques to the individual students. Some just amazed me with their work and ingenuity. Hands off was fine for such students. They reported to the class what they did or learned every week. Everybody wins. A few students, perhaps one to three from a course of ten to twelve, needed more assistance. The trick was to find the balance between doing too much or too little when guiding them.

I tried to work together with students on all aspects of their learning. For example, I felt is was important that every student know how to take a filament off the shelf and follow the printer's instruction to feed it into the printer. Teaching one or two students a task, as another example, how to balance the build plate, sufficed. They then passed the knowledge onto other students. Students became fluent in whichever software package they used to create their models. Sometimes students had ideas outside my expertise and I learned from them. In one of the courses students did some projects on a CNC drill press down the hall to "3D print" parts of geometric surfaces. Similarly some students experimented with our Anycubic resin 3D printer (with cure machine) which uses a very different type of 3D printing modality than the more standard extrusion 3D printers. Curing is a necessary step to get it to solidify. Resin is designed to be UV reactive, so exposing it to UV through a resin curing light-box or natural sunlight makes the print harder, stronger, and more durable.

My goal in these courses and projects was to have students master as many facets of 3D printing basics as possible, while focusing on one or two projects. In the classroom, students share with their fellow students their successes and failures. They pass along what they learned. Perhaps, a certain temperature setting worked better when using lots of support material. Or, a certain command in Mathematica was better for producing the desired output.

There were always issues, but the two that stand out in my memory are that the 3D printers regularly broke down and some types of filaments can get soggy due to humidity, causing bad extrusion and failed prints. When the course was taught with only one printer, when it broke down, I would spend hours fixing it. When it completely broke down we needed to send it back to Ultimaker to have it repaired. We worked on "3D Printing Theory" in the meantime. When we were able to purchase first another and then many more printers this became more of an annoyance than a show-stopper.

Roughly speaking there are three steps to 3D printing.

- Find or create an STL file of what you want to print. This typically involves a software package for model creation. Some of the famous ones (there are at least 40) are Tinkercad, Blender, and Mathematica. Some are free, some are not. Every model creation software has it own way of doing things. From the drag and paste Tinkercad to the more programming centered Mathematica. The choice of software very much depends on your goals.
- Load this STL file into a slicing software package that is compatible with the printer you'll use. At this point a multitude of settings may be selected. Every slicer has numerous settings. They involve type of filament, nozzle size, printer speed, printer precision, layer size, density, etc. This produces a printer specific file. For example the Ultimaker slicer (called Cura) produces a GCODE file. The Markforged slicer (called Eiger) produces a MFG file.
- Load this file into the printer, here again another multitude of settings may be selected, and print.

3. Preparing the course and materials

Before teaching a 3D printing class or mentoring a student 3D printing project you'll need a 3D printer. Typically this requires securing funds and space. Then selecting, ordering, unpacking, setting up, and testing the printer. This is a minimum, you'll also need filaments and tools. There may also be institutional health protocols to follow, such as wearing masks, etc.

There are many types of 3D printers. Extrusion printers (sometime called FDM or FFF printers) are the most common. Some famous extrusion printers (there are at least 20) include Ultimakers and Markforged. Cheaper extrusion printers that are used in teaching 3D printing courses in high schools are in my opinion not precise enough for most mathematically related projects. Filaments come wrapped around spools which are attached to the printer. The filaments are fed into a tube which guides the filament to the printer nozzle. During the printing process the nozzle become extremely hot and melts the filament and the melted filament is put, layer by layer, onto the printer bed where it cools into a hard object. Of the six printers currently in our lab, five are extrusion printers and one is a resin printer. There are a wide variety of other types of 3D printers, however I'll discuss only extrusion printers.

For every printer there is a list of filaments which will work on that printer. It is fair to say at this point that anything can be used as filament. The most common filament is plastic; with numerous types with various specific properties. Other common filaments include carbon fiber, metal, cement, food, skin, etc. Filaments are often referred to by nicknames, or acronyms such as Onyx or PLA. Over the years we've experimented with a wide variety of filaments, including magnetic and glow-in-the-dark. A very special type of "wash-away" (water soluble) filament will be discussed below.

One of the most important concepts in 3D printer is rebar or support material. If you are printing an umbrella (with the canopy on top) then if you've no support you'll eventually be printing in midair. Until the filament hardens something needs to temporarily hold it up. This is called support material (or rebar). After the 3D print is finished the rebar must be removed. This is often a long tedious process and can easily lead to damaged prints. There is an art to using enough, but not too much support for a print. Many of the slicer settings are related to this issue, such as when to introduce support and at what angle. The density setting for the support material is typically lower than for the print material.

Our first 3D printer had only one extruder (or nozzle). Therefore the support material was the same material used for the print itself. Some prints, for example the arms of the dancer below, required extensive support, and students might spend long periods of time removing the rebar without damaging the print.

Our second 3D printer had dual extrusion. It was therefore possible to use water soluble material for the support material. When the print was finished, it would be put in a bucket of water and by the next day all the rebar will have disappeared.

Because of our experience with 3D printers breaking down, admittedly after significant usage, reliability has become the number one consideration when purchasing a 3D printer. Other important factors to consider, depending on the goals, are the stiffness, strength and precision which the 3D printer is capable of. Naturally all considerations must include the cost of the printer and cost of the types of filaments it uses. Finally, simplicity and user friendliness are important.

4. Projects

Ian: Ian found his problem on the internet [**Wi**]. We agreed it was fascinating and might be fun to investigate with the help of 3D printing. The question is: How many unit tetrahedra (edge lengths all one) can be packed inside a unit sphere? This is an open question. On the one hand, using icosahedral symmetry, it is easy to squeeze in 20. On the other hand the volume of 23 such tetrahedra is greater than the volume of the unit sphere; so 23 is impossible. So the open questions are: Is 21 possible? Is 22 possible?

Ian attacked the problem in several ways. First he 3D printed the problem. See Figure 2. Next, noting that the question is equivalent to tiling the unit sphere with unit spherical triangles, he wrote a Mathematica program which would allow him to move as many as 22 such triangles as in Figure 3. He also 3D printed unit spherical triangles and investigated magnetic filament to stick these onto an actual unit sphere.

At the end of the semester, Ian continued working on the problem as a Senior Project, mostly with paper and pencil. Ian won first prize at our local MAA Sectional Meeting for the presentation of his many partial results. Ian is now working toward his PhD in Mathematics at the University of Colorado, Boulder.

A nice description of Ian's project was included in Scott Hershberger's review of the 2022 AMS Short Course on 3D Printing which recently appeared in the Notices of the AMS [**H**].

Sarah: Sarah was a double major in Mathematics and Dance. She was interested in 3D printing something Dance related. After trial and error we landed on the idea of an ellipsoidal dancer in various poses. She used Mathematica to

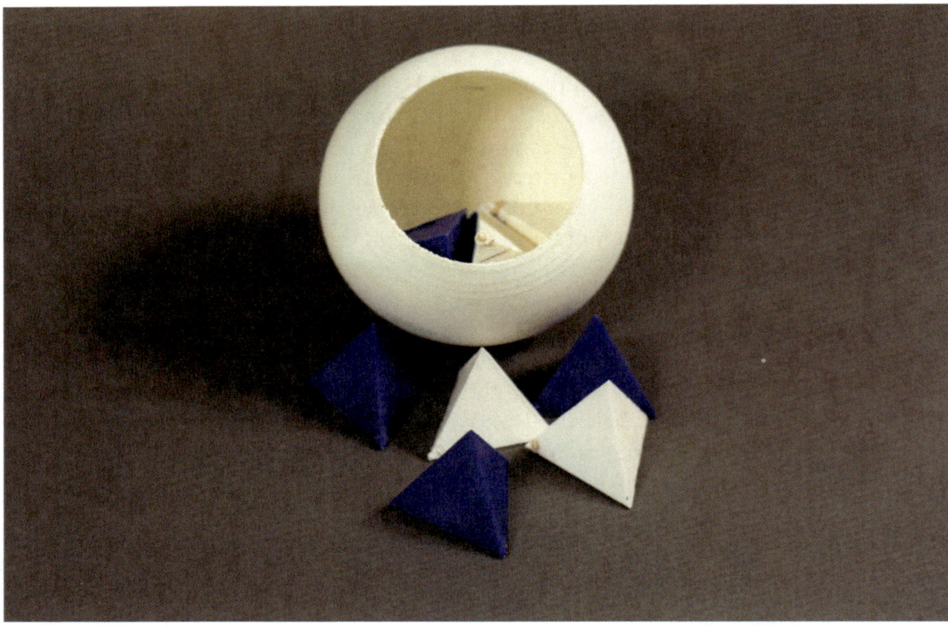

FIGURE 2. Packing equilateral unit tetrahedra into a unit sphere

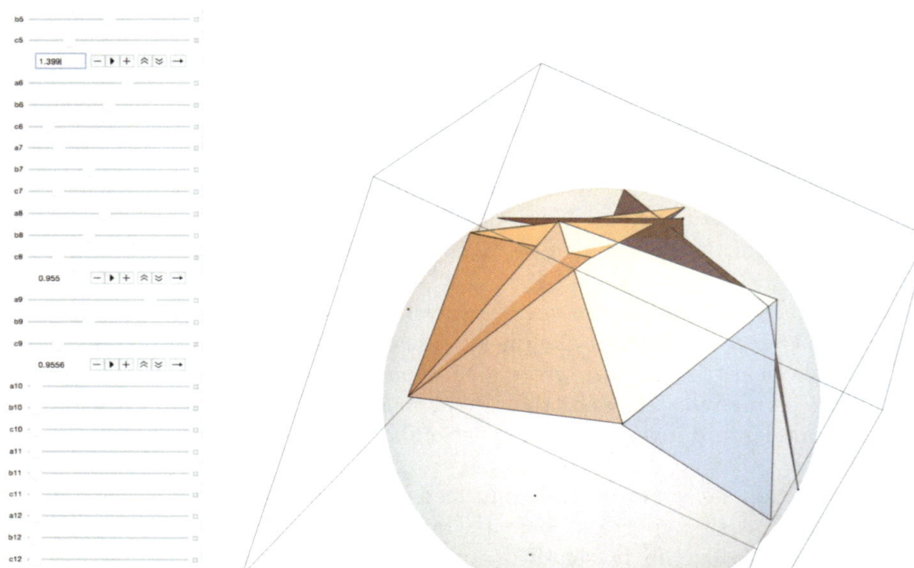

FIGURE 3. Using Mathematica to find optimal packing

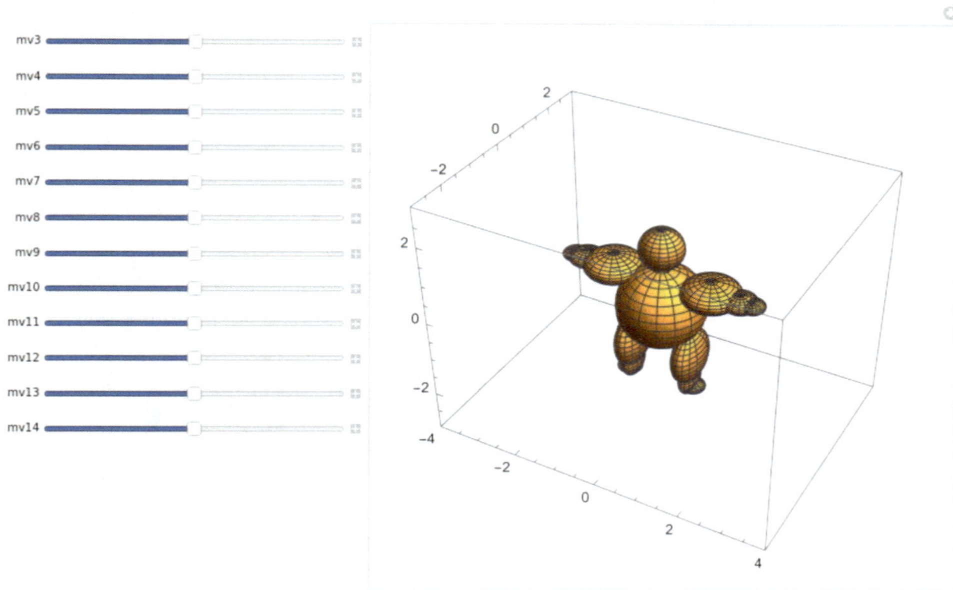

FIGURE 4. Using Mathematica to find perfect pose

FIGURE 5. Ellipsoid dancer striking a pose

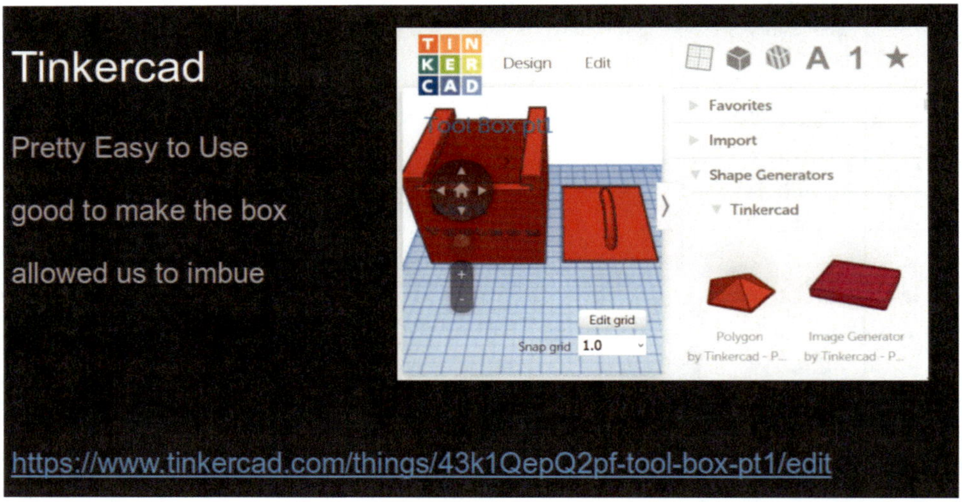

FIGURE 6. Tinkercad toolbox

FIGURE 7. The shell method

generate her STL file. Head, body, arms, etc. were all made out of small, large, pancake, cigar, etc. ellipsoids. Using the Mathematica Manipulate command, Figure 4, Sarah was able to consider an infinite variety of poses. The most fun was setting many (or even all) slider on animate and to watch the crazy ellipsoidal dancer wildly dancing. The final pose is seen in Figure 5.

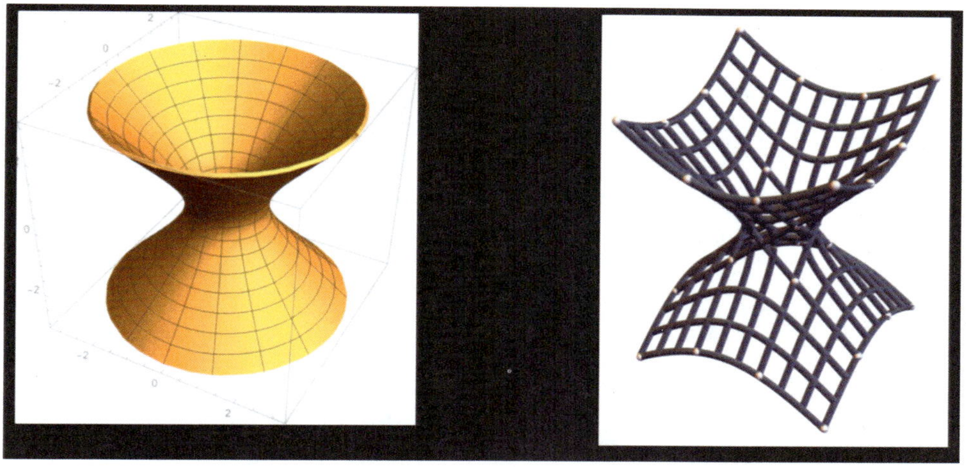

FIGURE 8. Hyperboloids for vector calculus

James and Robert: James and Robert made a nice toolbox using Tinkercad, see Figure 6. They also 3D printed a wide variety of curves and surfaces. These included 3D models that demonstrate concepts from Calculus, such as the shell method, see Figure 7, and from Vector Calculus, such as quadric surfaces. They printed some surfaces, for example the Hyperboloid in Figure 8 using surface modeling and wire framing modeling.

Maxime and Rose: Coincidentally, just before the start of my Fall 2018 course, I was contacted by one of my former colleagues at the University of Toledo. He has been the point person in creating a "Bronze artwork of a three-lobed Wente Torus to be displayed on UT's main campus for campus beautification." The Wente Torus is a famous soap bubble surface (see definition below) discovered in 1984 by Henry Wente, who was a mathematics professor at the University of Toledo. Wente found the first soap bubble that was not the round sphere, solving a long-standing open problem in geometry [**PS89**]. Henry passed away January 20, 2020.

Maxime and Rose loved that story and the idea. They worked, using Mathematica, on solving the differential equations with difficult closing conditions and succeeded in making a very nice STL file with tubular coordinates. These coordinates are called curvature line coordinates. Holding the surface in your hands allows you to better see the amazing property that one family of coordinates always lie in planes and that the other family of coordinates always lie in spheres. See Figure 9. We sent it to Toledo where it was fabricated by the Department of Art Center for Sculptural Studies.

Geometric fabrics - entire class fall 2021: Both the Wente Torus and the surfaces discussed below are examples of surfaces immersed into $\mathbb{R}^3$. Our goal for the Fall 2021 was to make dynamic 3D versions of such surfaces which will lead to completely new types of fabrics, which we call "geometric fabrics". Only a handful of research groups in the world are working on making prototypes of such surfaces as the necessary theory has only recently been developed. This made the course both exciting and foreboding. We begin with a short overview of terms from

FIGURE 9. Wente torus based on Maxime and Rose's code 2018

classical surface theory. The definition of these surfaces, in the hyperbolic case, is given below.

For surfaces immersed into $\mathbb{R}^3$ there are two main invariants: Gauss curvature K and mean curvature H. Surface theory focuses mainly on the cases when either K or H are constant along the entire surface. In particular minimal surfaces have $H \equiv 0$ and soap bubbles have $H \equiv 1$. Also flat surfaces have $K \equiv 0$ and finally $K \equiv -1$ surfaces are called hyperbolic. These invariants are similar to the the two invariants of a curve which students learn in Vector Calculus: the curvature κ and the torsion τ (which measures twisting) of a curve in $\mathbb{R}^3$. The definitions of geometric fabrics precisely discretizes local curvature conditions. There are no hinges in these models.

Our surfaces follow very specific rules and differ from the many famous fabrics which were created with completely different discretizations. These include both Thurston-Henderson-Tiaminov's wonderful knitted hyperbolic surfaces [**HT01**], and Segerman's brilliant hinged hyperbolic surfaces [**Se**]. The surface in Figure 10 is

FIGURE 10. Wunderlich's hyperbolic fabrication 1951

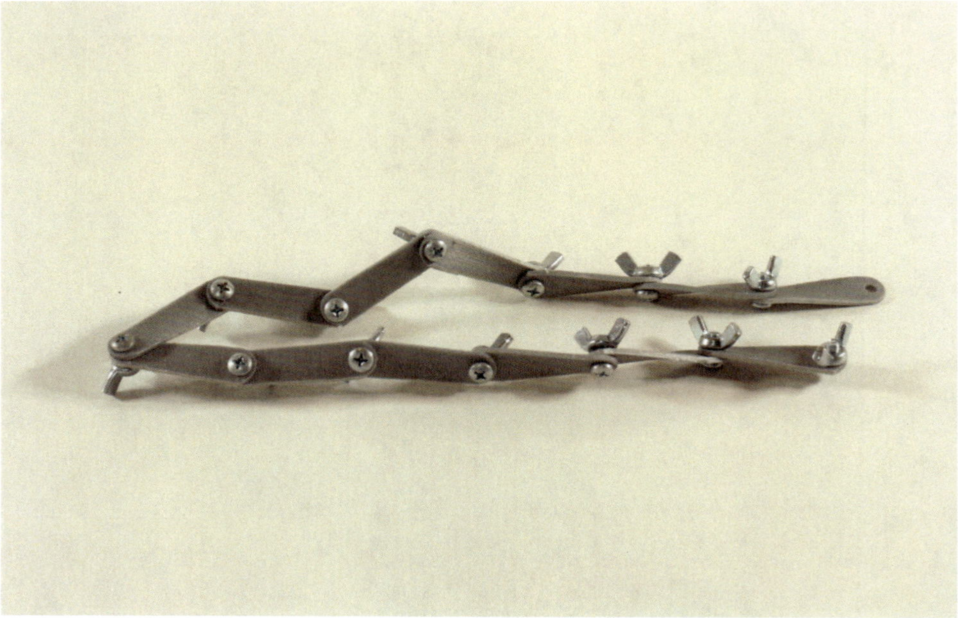

FIGURE 11. Discrete curve of constant torsion

not movable, so we don't consider it a fabric. For minimal surfaces, soap bubbles, flat and hyperbolic surfaces there corresponds a geometric fabric. The fabrication processes are based on the well established, but quite specialized theory of the discrete differential geometry of surfaces. If you follow the rules for that type your fabric will always result in discrete versions of that type. In the class we focused almost exclusively on the hyperbolic case.

Figure 10 shows a non-movable fabricated hyperbolic surface made by Wunderlich, [**Wun51**], in 1951. The coordinate curves are discrete curves of constant torsion discussed below. Our goal was to 3D print movable fabricated versions of this surface. I lectured regularly during the course on the basics of surfaces theory, especially hyperbolic surfaces and their mathematical hyperbolic discretizations. Such a fabric will, by construction, always be hyperbolic. Furthermore, every hyperbolic surface can be tightly dressed by this fabric. I want to emphasize that these 3D prints are prototypes only and are at least ten times larger than the actual fabric will be. The theory of discrete curves and surfaces is beyond the scope of this article, so we will focus on only the main concepts without formulas. The notion of torsion one learns in vector calculus is assumed.

A discrete curve is a sequence of vertices. Roughly speaking the discrete curvature of a curve at a vertex is determined by the angle between the edges which meet there. The discrete torsion of a edge is determined by how much it twists. A bike chain is the fabrication of a discrete curve with constant zero torsion with arbitrary curvatures. Our interest is in discrete curves with constant non-zero torsion (each edge needs to be twisted by the same amount) and still with arbitrary curvatures. See Figures 11 and 12.

FIGURE 12. Another curve of constant torsion

Before discussing the more difficult case of discrete hyperbolic surfaces, let's consider the simpler case of a discrete planar surfaces. Some examples guided the students to understanding this type of surface: (Typically older) outer elevator doors, folding garden fences, or (an imagined) planer web of bike chains;one family of chains horizontal and one family vertical. See Figures 13 and 14 for such a fence.

Through the history of surface theory there have been attempts of discretize surfaces in a way that preserves the fundamental invariants such as mean and Gauss curvature. One of the early examples of this is in the case of discrete hyperbolic surfaces. In Bennett's 1914 foundational paper, [**B**], he breaks down the rules for the basic parts. He calls these *skew isogram mechanisms*. The term "isogram" is used to describe any four-sided figure, which has its alternate sides equal in length. It was well known at the time, that certain special coordinate curves on hyperbolic surfaces are always the constant torsion as curves in $\mathbb{R}^3$ mentioned above. This guided Bennett into requiring his discrete coordinate curves to have constant torsion, as discussed above. In the thought experiment of the the web of bike chains mentioned for the flat case we need to imagine the links of the horizontal (resp. vertical) bike chains will be twisted by the same positive (resp. negative) amount. See again Figure 10. The simplest version of Bennett's Rules are

- Each vertex has four edges.
- All edges have length one.
- The four edges at an vertex are coplanar.

See Figures 15 and 16 for a prototype hyperbolic fabric.

If one studies Figures 15 and 16 one can begin to see one of our most difficult challenges: the rivets. The rivets need to be tight enough to keep their four edges perfectly co-planar; but no so tight that the edges can't rotate. They need to be

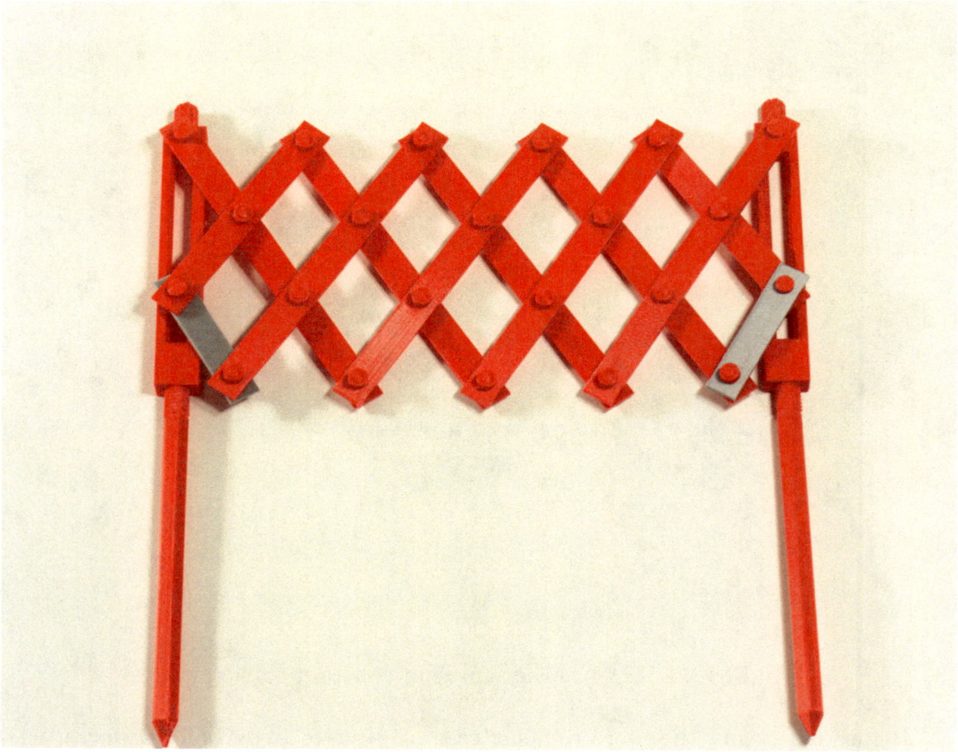

FIGURE 13. Red fence open

strong enough to not break; but small enough to make the piece of fabric reasonably sized. This is an ongoing process.

Related long standing open problem: There is a simple (to state) open problem in the surface theory of immersions into $\mathbb{R}^3$: Does there exist a hyperbolic Mobius strip smoothly immersed into $\mathbb{R}^3$? [**N**]. It is well known that there are flat and minimal Mobius strips; and that there are no soap bubble Mobius strips. So why not 3D print some hyperbolic paper and try to build one. That was the 2016 project of two students. Using a variety of filaments, thicknesses, and sizes they printed out a piece of the standard hyperbolic plane known as the pseudo-sphere. Then they cut it into strips and attempted to twist the strips into Mobius strips. In the end it was inconclusive. When such a strip was formed, it was unclear if it was maintaining the the hyperbolic condition of constant curvature. The situation is similar to the more familiar flat piece of paper case. When one twists a flat piece of paper into a Mobius strip, it is no longer flat. To prove the existence of an everywhere flat Mobius strip immersed into $\mathbb{R}^3$ is not a trivial exercise and was first demonstrated by Sadowsky, [**S**], in 1930. His proof uses generalized cylinders and the proof doesn't carry over to the hyperbolic case. It is hoped that better prototype pieces of discrete hyperbolic patches can be used to make discrete hyperbolic Mobius strips, or to indicate why its not possible. We are currently working on making finer and stiffer prototypes, keeping this open math question in mind.

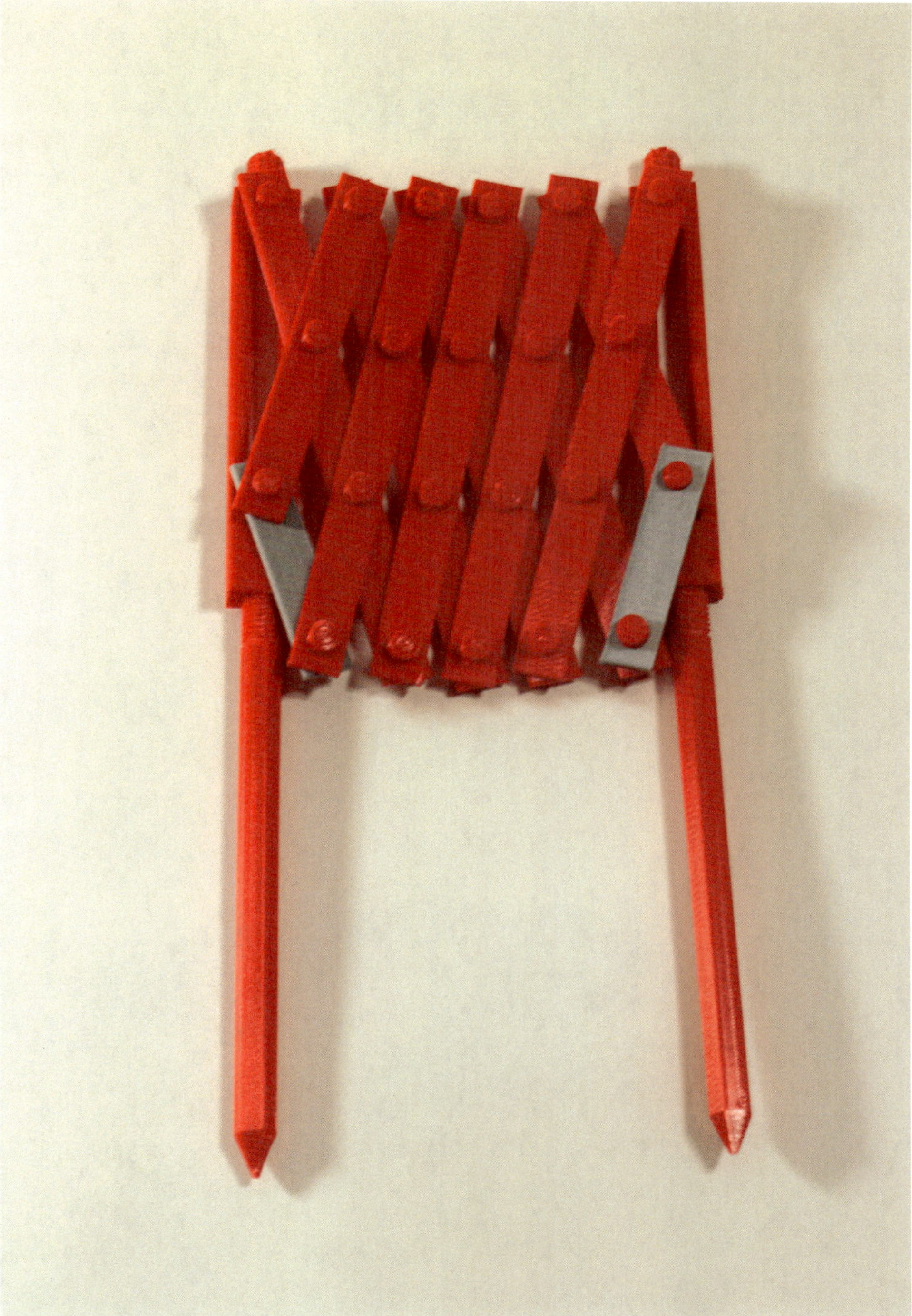

FIGURE 14. Red fence closed

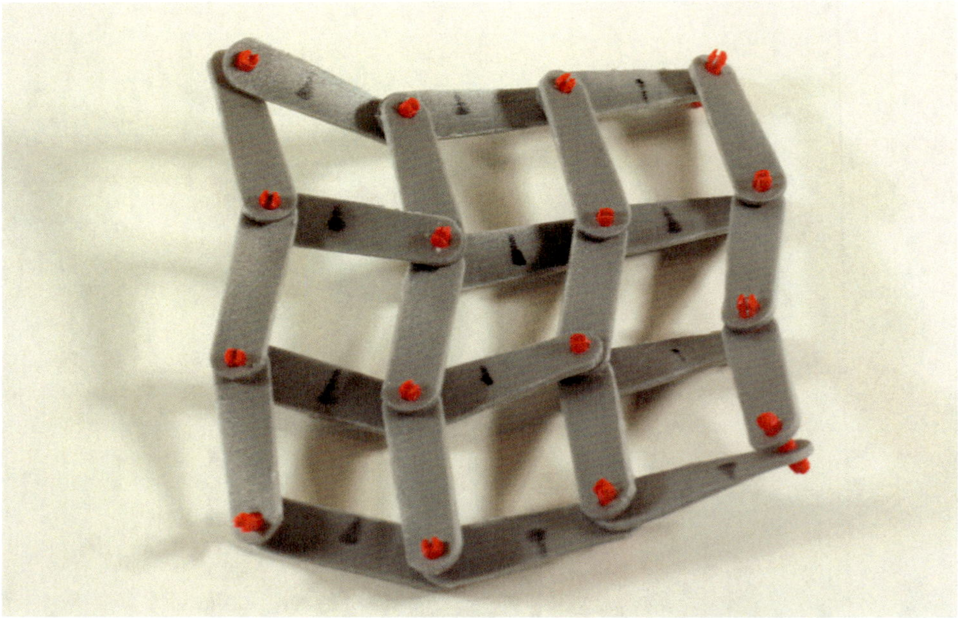

FIGURE 15. Hyperbolic fence open

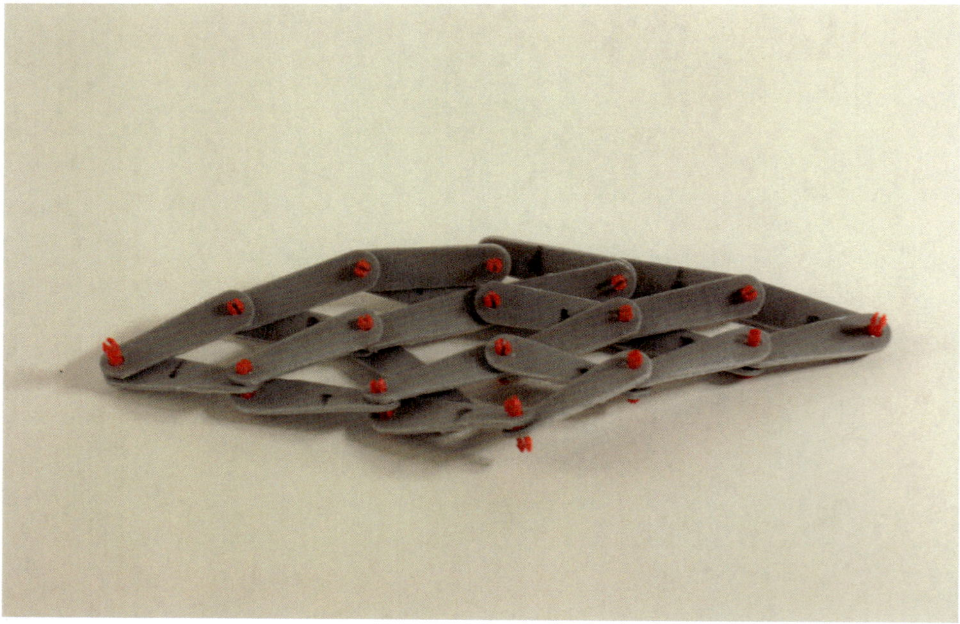

FIGURE 16. Hyperbolic fence closed

5. The future of 3D printing at St Mary's College of Maryland

We have worked with many departments on campus and in the local community. The Art Department added two 3D printers to their lab in the Fall of 2021 and are in the the process of integrating them into their curriculum. Our Chemistry Department has an excellent 3D printing lab led by Troy Townsend. After David Gerrish, a student in my 2016 class, finished his work with me on mathematical surfaces, he joined Troy's team. In particular David was always interested in the mechanics of 3D printers and his Senior Project, with Troy, consisted of building a specialized printer. David, after graduating, started is own 3D printing company. Troy has sent several chemistry students to take my 3D classes to gain some experience. Most recently students took my Fall 2021 course before joining Troy's research group which is 3D printing solar panels. The summer after the COVID-19 pandemic started Troy borrowed all the 3D printers on campus, including the two we currently had in our department. With a team of students his lab produced, 24-7, ventilators for the St. Mary's County Hospital.

Over the years our department has 3D printed a wide variety of one-off items for several departments on campus.

St. Mary's College of Maryland was awarded a significant four-year Sherman Fairchild Foundation grant focused on the creation of a fabrication center for use between several departments including our Mathematics and Computer Science Department. In June 2022, we purchased an expensive Markforged 3D printer which is capable of making smaller, stronger, and stiffer metal-like parts which should drastically improve the geometric fabric prototypes discussed above.

References

[B] G. T. Bennett, *The Skew Isogram Mechanism*, Proc. London Math. Soc. (2) **13** (1914), 151–173, DOI 10.1112/plms/s2-13.1.151. MR1577497

[H] S. Hershberger, *From Barth Sextics to Minimal Surfaces: Printing Mathematics in 3 Dimensions*, Notices of the AMS, (May 2022), pp. 861–865.

[N] Kosuke Naokawa, *Negatively curved Möbius strips on given knots*, J. Knot Theory Ramifications **22** (2013), no. 7, 1350023, 12, DOI 10.1142/S0218216513500235. MR3084747

[PS89] U. Pinkall and I. Sterling, *On the classification of constant mean curvature tori*, Ann. of Math. (2) **130** (1989), no. 2, 407–451, DOI 10.2307/1971425. MR1014929

[S] W. M. Sadowsky, *Ein elementarer Beweis fur die Existenz eines abwickelbaren Möbiusschen Bandes und Zuriickfiihrung des geometrischen Problems auf ein Variationsproblem*, Sitzungsber. Preuss. Akad. Wiss., 22 (1930) pp. 412–415.

[Se] H. Segerman, *Hinged negatively curved surfaces*, `youtube.com/watch?v=RnKuIbKauXk&ab_channel=HenrySegerman`, (2014).

[HT01] David W. Henderson and Daina Taimina, *Crocheting the hyperbolic plane*, Math. Intelligencer **23** (2001), no. 2, 17–28, DOI 10.1007/BF03026623. MR1843235

[Wi] I. O. Wilhelm *Packing triangles on a sphere*, `www.math.uchicago.edu/$~$may/VIGRE\/VIGRE2008/REUPapers/Wilhelm.pdf`.

[Wun51] Walter Wunderlich, *Zur Differenzengeometrie der Flächen konstanter negativer Krümmung* (German), Österreich. Akad. Wiss. Math.-Nat. Kl. S.-B. IIa **160** (1951), 39–77. MR56342

Mathematics and Computer Science Department, St. Mary's College of Maryland, St. Mary's City, Maryland, 20686

Email address: `isterling@smcm.edu`

Proceedings of Symposia in Applied Mathematics
Volume **79**, 2023
https://doi.org/10.1090/psapm/079/00717

Some mathematical problems motivated by 3D printing

Maria Trnkova and Andrew Yarmola

ABSTRACT. 3D printing is a rapidly developing technology in industry with deep connections to engineering, mathematics, and computer sciences. This survey will describe the mathematical nature of several problems arising from 3D model design, printing and 4D printing, while providing resources for work done in these directions. It might be of interest to mathematicians looking for new research projects.

1. Introduction

3D printing is a relatively new branch of engineering which faces many challenges, mainly mechanical, material, design, computational and algorithmic aspects. The process of printing can be divided roughly into three steps: model design and storing; preparation for printing via slicing and infill algorithms; and physical process of layer deposition in the printer. An additional fourth step could be the considerations of the dynamics properties of the object, often called 4D printing. Each step brings its own challenges to different areas of science, including mathematics. Problems arising from each of these steps admit for a wide range of solutions, each particular one depending on the nature of the model and its application. The application spectrum is wide and keeps expanding: from engineering to automotive and aerospace, from medicine and biological systems to food industry, etc. There is no one best solution and only a balance between different needs can be achieved. Therefore, the list of open problems is never ending. In this expository article we try to discuss some of these challenges, give the current status of each problem and mention some novel approaches. A good survey on the status of 3D printing and its engineering challenges is given in [**GZR$^+$15**].

2. Model design and Storing

Many aspects of model design can be considered mathematical problems. Additive manufacturing allows one to create very complicated geometries but there are some factors which have to be considered when designing the geometric structure. These include gravity (center of mass), forces that will be applied to the object (load), robustness and deformation. For considerations when using elastic material for printing, see [**LSWW14**]. The main constraints when designing a model come from the later stages of printing and are described below.

Models can be described algebraically but they are usually stored as a 3D mesh, which is later "sliced" for printing. Meshes for large and complicated models with fine details can result in a huge number of small triangles or polygons, therefore the data storage becomes an issue. Finding efficient and optimized representations of models is still an active area of research. For example, recent advances in graphics hardware and algorithms have allowed for non-uniform rational basis spline (NURBS) surfaces to enter the mainstream. Some widely used file formats like STL or OBJ are sensitive to errors and do not always preserve the correct topology of the surface (not "water-tight"). The most recent alternative approaches offer volumetric representation (voxel-based) to approximate the solid object and also to store metadata of a model such as scaling and layering, multi-material usage, copyrights, etc. Such files can directly use scanned images from MRI or CT scans but the main disadvantage here is a huge memory consumption, see [**TSH98**, **RMvTK**$^+$**10**, **YW13**].

3. Preparation

The process of preparing a 3D model for printing can be quite involved and heavily depends on the printing technology. Preparation can be optimized in several ways by different parameters. A basic parameter is orientation: a choice of the best position of the model such that the print requires less support, remains stable during printing, and returns a smooth surface. There is also an endless number of problems related to the best print pattern for filling internal space in an object driven by different requirements, such as density, rigidity, and other mechanical properties.

Support. For extrusion printers, optimal construction of support for protruding parts with overhangs and undercuts is a big area of interest. Most of the time overhangs are addressed by building large support structures which have to be carefully removed after printing. The larger the contact area between support material and the model, the higher the failure rate and the rougher the print. If the support uses soluble material then it can be washed away. Non-soluble material can be removed mechanically using pliers or carving/milling and then sanding and polishing or vapor smoothing. See, for example, [**UULH21**]. While many support-generating algorithms are already included in 3D slicing software, research on new and better techniques is ongoing. There are also optional approaches, such as wall thickening, which completely avoid supports or significantly reduce the support structure [**AD95**]. Again, the challenges are motivated by the nature of the requirements: the need to save material, reduce waste and printing time, to provide rigid support but make sure it is removable and leaves a smooth surface. Another approach to address these difficulties is to break up the model into pieces that will be assembled later. This will be considered below.

Problem: *Create a support-generating algorithm for 3D models that balances optimal model rotation, minimum waste design, material cost, printing time, surface quality of the final object, sagging of overhanging parts, invisibility of traces from support removal and low density of support.* It is also a good idea to keep in mind the potential user's input (from experienced to novel users).

Model decomposition. Some of the challenges mentioned above can be resolved by decomposing a model into parts which will be assembled after printing. Model decomposition can solve the support problem if the model is cut and oriented in such way that its pieces do not require any or only minimal support. Additionally, sometimes it is necessary to decompose a model into pieces such that they can pack the smallest space. For example, such packing is essential for powder bed fusion printers when the model is too large for a printing box or is too sparse to be cost effective.

Problem: *Optimize a model decomposition by the number of cuts and quantity/quality of support required for each part. Orient pieces such that the most visible parts of the assembled surface are smooth after printing.*

Another variation of this problem:

Problem: *Decompose a model into pieces which will pack the given box as tight as possible.*

Some progress in these directions can be found in [**WDF**$^+$**20**, **Att15**].

Filling. Since many models are provided as oriented surfaces, it is often necessary to fill the interior structure. One may also think of this infill as "internal" support. Often the quality and detail of this structure does not need to be as fine as the surface and therefore speed and material concerns are most prevalent. There exist many different infill patterns (i.e. nozzle paths for infill) each having their own properties (rectangular, square, triangle, wiggle, criss-cross, honeycomb, gyroid, etc.). They can be characterized by model strength, printing speed, the amount of material used, density, weight, etc. For example, it might be desirable to have homogeneous infill which looks the same up-down as it does side-to-side. Orientation of the infill could be important when moving to the next step of printing– slicing– as we would like to avoid horizontal or low angled walls of the infill. Some models may require rigidity while the other has to be flexible or elastic. One may care about the topology of the infill structure: connected or isolated cells. Removal of the infill, once printing is completed, requires the use another material; also a design should allow soaking of the interior, see [**Raj22**].

In medical applications, soft tissues and their internal structure is a big concern. For example, silicone phantoms mimic imaging properties of soft tissues and can be used to study radiological properties for CT scans. They are made with multi-material extrusion 3D printing technology. Different density is used in fabrication of imaging phantoms as it affects their radiological properties [**HJK**$^+$**22**].

Problem: *Create a new infill pattern which optimizes some of the above mentioned or other properties.*

For example, novel solutions to the problem, focused on a good balance of rigidity and weight, is achieved by bone-like structures [**WAWS18**]. An additional area of consideration would be the acoustic properties of infill patterns. In particular, this might be an issue for medical devices, such as hearing aids, and structurally important prints such as load bearing parts or large-scale printing such as buildings.

4. Printing

Once a model is ready for printing, it is sliced into horizontal layers. The thickness of each layer and, in the extrusion setting, the nozzle path/trace, are important considerations. Each step can be performed in different ways bringing their own challenges.

Slicing. Printing will be faster when layers are thick, but this will lead to the "stair-stepping" effect of the printed model surface, model weakness, and other errors [**ZC09**, **HWC13**]. Thinnest layers, as low as ~ 0.02 mm, can be obtained by material jetting systems. The thickest layers are used for printing buildings and large structures. Highest quality extrusion prints are often obtained at slow nozzle speeds and thin layers, but this comes at the significant cost of increased printing time. Further, the printing processes itself is prone to error. Layers may be misaligned with the geometry of the model, leading to incorrect topology and the splitting of layers. A way of correcting these issues, before and during the printing process, is an active field of research. Some approaches use adaptive slicing – a combination of thin and thick layers – others include sanding the final product and vision guided realignment during printing [**WHZ17, PZZC12**].

Problem: *Improve algorithms of approximating a 3D object by horizontal slices preserving the topology of the object and keeping its geometry as close to the original one as much as possible. Create/improve visual feedback realignment algorithms.*

Tracing. A 3D printer prints one layer at a time. A layer may look like a union of disconnected, often not simply connected, regions. Further, in the extrusion setting, the nozzle size determines the minimal horizontal width of a layer feature. Mathematically, one can think of the nozzle path as a collection of thickened non-overlapping curves whose union is the layer. Examples of algorithms that generate such paths include mathematical space-filling curves. Extra challenges arise from the physical properties of the nozzle. While the perfect nozzle can make jumps between components of a disconnected layer, there is usually a problem called "stringing" where bits of material are still extruded during such jumps. Shorting the quantity and distance of such jumps is considered beneficial. Additionally, temperature distribution could also play a significant role in printing. If a model is evenly heated during the printing process it returns a better bond between layers. Therefore, we would like to trace the region without keeping the nozzle at the same position for a long time. Other restrictions to consider are crossover of the path and moving the nozzle along sharp corners. See [**GKD20, ZGH$^+$16**].

Problem: *Create an algorithm for tracing a set of disconnected and not simply connected regions with the restrictions mentioned above.*

Advanced 3D printing. Printing does not need to start necessarily on a flat surface. Printing can be done on a non-flat printing beds as well. There is some work demonstrating 3D printing on a tessellated surface. The main challenge here is to project a single layer onto the given surface. Since a straight line segment will project onto a union of line segments bent along surface edges, different projections can be used to find the projection of a layer. The best reference to this work can be found here [**RPCURF$^+$21**]. There is also research into 6-axis 3D printing nozzles which can tilt and print at an angle. Such nozzles allow for printing from different directions, so often the prints do no require supports. Working with such nozzles and designing tracing algorithms is an ongoing area of research, see for example [**OK22**].

5. 4D printing

Another challenge is to print objects whose physical properties change after manufacture. This is known as 4D printing. It can include the dynamical properties of models such as hinges or joints, for example a Rubik's cube, or shape-morphing objects.

Shape morphing. 3D printing can be done not just with rigid materials. Variations of the material, its density, and print pattern can produce objects whose final shape depends on the physical environment. For example, cellulose fibrils used in hydrogel composite ink swell when immersed in water. A model printing with such fibrils is forced to warp and twists when water is introduced. This changes the Gaussian curvature of the model in realtime. Depending on the layering of the model it is possible to predict some aspects of this behavior, see [**SGMN$^+$16**]. Such work raises a purely mathematical question:

Problem: *Consider two layers of tubes projecting to a union of curves in a simply connected domain R in $\mathbb{R}^2$. Assume each tube has a fixed stretching factor. Is there always a way to attach tubes to R as above such that after their stretching the domain is isometric to a given simply connected surface embedded in $\mathbb{R}^3$?*

Partial answer to this question is given by [**ASK14**], where they connect the nematic director field of the region with the intrinsic geometry of the constant curvature surfaces. Their paper has many references to other similar works, see [**CL14, SCL02, MBW11**].

Dynamical models. Some models have the ability to move their parts after being printed. One of the advantages of additive manufacturing over subtractive processes is that moving pieces can be printed together and often do not require assembly. For example, a triple gear model can be printed as one piece but has movable parts [**SS13**]. Other models can be printed in parts and then assembled into one dynamical model, such as scissors grids [**SV22**]. Usually, such models require a gear mechanism adjusted for 3D printing, see [**MS19, MSE20**] and the first chapter of this volume.

Puzzles. Entertainment and puzzles are also a valuable aspects of 3D printing. There is a wide horizon of puzzles and games that have yet to be created. For readers interested in readily available models, we suggest Henry Segerman's, George Hart's, and Oskar van Deventer's websites: for example, 15+4 or coiled puzzle frame, holonomy mazes, quintessence puzzles, see [**Seg, Har, vD**].

Problem: *Create a new puzzle of your own.*

References

[AD95] Seth Allen and Deba Dutta, *Determination and evaluation of support structures in layered manufacturing*, Journal of Design and Manufacturing **5** (1995), 153–162.

[ASK14] Hillel Aharoni, Eran Sharon, and Raz Kupferman, *Geometry of Thin Nematic Elastomer Sheets*, Physical Review Letters **113** (2014), no. 25, 257801 (en).

[Att15] Marco Attene, *Shapes In a Box: Disassembling 3D Objects for Efficient Packing and Fabrication: Shapes In a Box*, Computer Graphics Forum **34** (2015), no. 8, 64–76 (en).

[CL14] Brett G. Compton and Jennifer A. Lewis, *3D-Printing of Lightweight Cellular Composites*, Advanced Materials **26** (2014), no. 34, 5930–5935 (en).

[GKD20] Prashant Gupta, Bala Krishnamoorthy, and Gregory Dreifus, *Continuous tool-path planning in a graphical framework for sparse infill additive manufacturing*, Comput.-Aided Des. **127** (2020), 102880, 16, DOI 10.1016/j.cad.2020.102880. MR4099852

[GZR$^+$15] Wei Gao, Yunbo Zhang, Devarajan Ramanujan, Karthik Ramani, Yong Chen, Christopher B. Williams, Charlie C.L. Wang, Yung C. Shin, Song Zhang, and Pablo D. Zavattieri, *The status, challenges, and future of additive manufacturing in engineering*, Computer-Aided Design **69** (2015), 65–89 (en).

[Har] George Hart, *George w. hart — index*, `http://www.georgehart.com`.

[HJK+22] Sepideh Hatamikia, Laszlo Jaksa, Gernot Kronreif, Wolfgang Birkfellner, Joachim Kettenbach, Martin Buschmann, and Andrea Lorenz, *Silicone phantoms fabricated with multi-material extrusion 3d printing technology mimicking imaging properties of soft tissues in ct*, `arXiv:2212.03116` [physics] (2022).

[HWC13] Pu Huang, Charlie CL Wang, and Yong Chen, *Algorithms for layered manufacturing in image space*, Ph.D. thesis, Chinese University of Hong Kong, 2013.

[LSWW14] Ligang Liu, Ariel Shamir, Charlie Wang, and Emily Whitening, *3D printing oriented design: geometry and optimization*, SIGGRAPH Asia 2014 Courses on - SA '14 (Shenzhen, China), ACM Press, 2014, p. 1 (en).

[MBW11] C. D. Modes, K. Bhattacharya, and M. Warner, *Gaussian curvature from flat elastica sheets*, Proc. R. Soc. Lond. Ser. A Math. Phys. Eng. Sci. **467** (2011), no. 2128, 1121–1140, DOI 10.1098/rspa.2010.0352. MR2774553

[MS19] Elisabetta A. Matsumoto and Henry Segerman, *Geared jitterbugs*, Proceedings of Bridges 2019: Mathematics, Art, Music, Architecture, Education, Culture (Phoenix, Arizona) (Susan Goldstine, Douglas McKenna, and Kristóf Fenyvesi, eds.), Tessellations Publishing, 2019, pp. 399–402.

[MSE20] Jinesh Machchhar, Henry Segerman, and Gershon Elber, *Conjugate shape simplification via precise algebraic planar sweeps toward gear design*, Computers & Graphics **90** (2020), 1–10 (en).

[OK22] Sinan Olcun and Roger Kempers, *Thermal characterization of continuous pitch carbon fiber 3d-printed using a 6-axis robot arm*, 2022 21st IEEE Intersociety Conference on Thermal and Thermomechanical Phenomena in Electronic Systems (iTherm), 2022, pp. 1–6.

[PZZC12] Yayue Pan, Xuejin Zhao, Chi Zhou, and Yong Chen, *Smooth surface fabrication in mask projection based stereolithography*, Journal of Manufacturing Processes **14** (2012), no. 4, 460–470 (en).

[Raj22] Santosh Rajkumar, *Effect of infill pattern and build orientation on mechanical properties of fdm printed parts: An experimental modal analysis approach*, Preprint, `arXiv:2202.05692`, 2022.

[RMvTK+10] F. Rengier, A. Mehndiratta, H. von Tengg-Kobligk, C. M. Zechmann, R. Unterhinninghofen, H.-U. Kauczor, and F. L. Giesel, *3D printing based on imaging data: review of medical applications*, International Journal of Computer Assisted Radiology and Surgery **5** (2010), no. 4, 335–341 (en).

[RPCURF+21] Consuelo Rodriguez-Padilla, Enrique Cuan-Urquizo, Armando Roman-Flores, José L Gordillo, and Carlos Vázquez-Hurtado, *Algorithm for the conformal 3d printing on non-planar tessellated surfaces: Applicability in patterns and lattices*, Applied Sciences **11** (2021), no. 16, 7509.

[SCL02] James E. Smay, Joseph Cesarano, and Jennifer A. Lewis, *Colloidal Inks for Directed Assembly of 3-D Periodic Structures*, Langmuir **18** (2002), no. 14, 5429–5437 (en).

[Seg] Henry Segerman, *Henry segerman's webpages*, `http://www.segerman.org`.

[SGMN+16] A. Sydney Gladman, Elisabetta A. Matsumoto, Ralph G. Nuzzo, L. Mahadevan, and Jennifer A. Lewis, *Biomimetic 4D printing*, Nature Materials **15** (2016), no. 4, 413–418 (en).

[SS13] Saul Schleimer and Henry Segerman, *Triple gear*, Proceedings of Bridges 2013: Mathematics, Music, Art, Architecture, Culture (Phoenix, Arizona) (George W. Hart and Reza Sarhangi, eds.), Tessellations Publishing, 2013, pp. 353–360.

[SV22] Henry Segerman and Kyle VanDeventer, *Self-similar quadrilateral tilings and deployable scissor grids*, Proceedings of Bridges 2022: Mathematics, Art, Music, Architecture, Culture (Phoenix, Arizona) (David Reimann, Douglas Norton, and Eve Torrence, eds.), Tessellations Publishing, 2022, pp. 313–316.

[TSH98] U. Tiede, T. Schiemann, and K.H. Hohne, *High quality rendering of attributed volume data*, Proceedings Visualization '98 (Cat. No.98CB36276) (Research Triangle Park, NC, USA), IEEE, 1998, pp. 255–262,.

[UULH21] Erva Ulu, Nurcan Gecer Ulu, Jiahao Li, and Walter Hsiao, *Curvy: An Interactive Design Tool for Varying Density Support Structures*, February 2021, `arXiv:2102.10013` [cs].

[vD] Oskar van Deventer, *Puzzles by Oskar van Deventer*, `https://www.oskarvandeventer.nl`.

[WAWS18] Jun Wu, Niels Aage, Rudiger Westermann, and Ole Sigmund, *Infill optimization for additive manufacturing–approaching bone-like porous structures*, IEEE Transactions on Visualization and Computer Graphics **24** (2018), 1127–1140.

[WDF+20] Chenming Wu, Chengkai Dai, Guoxin Fang, Yong-Jin Liu, and Charlie C. L. Wang, *General Support-Effective Decomposition for Multi-Directional 3-D Printing*, IEEE Transactions on Automation Science and Engineering **17** (2020), no. 2, 599–610.

[WHZ17] Florens Wasserfall, Norman Hendrich, and Jianwei Zhang, *Adaptive slicing for the FDM process revisited*, 2017 13th IEEE Conference on Automation Science and Engineering (CASE) (Xi'an), IEEE, August 2017, pp. 49–54.

[YW13] Yuen-Shan Leung and C. C. L. Wang, *Conservative Sampling of Solids in Image Space*, IEEE Computer Graphics and Applications **33** (2013), no. 1, 32–43.

[ZC09] Chi Zhou and Yong Chen, *Three-dimensional digital halftoning for layered manufacturing based on droplets*, Transactions of the North American Manufacturing Research Institution of SME **37** (2009), 175–182.

[ZGH+16] Haisen Zhao, Fanglin Gu, Qi-Xing Huang, Jorge Garcia, Yong Chen, Changhe Tu, Bedrich Benes, Hao Zhang, Daniel Cohen-Or, and Baoquan Chen, *Connected fermat spirals for layered fabrication*, ACM Transactions on Graphics (TOG) **35** (2016), no. 4, 1–10.

Mathematical Sciences Building, One Shields Ave, University of California, Davis California 95616

Email address: `mtrnkova@math.ucdavis.edu`

Fine Hall, Washington Road Princeton New Jersey 08544-1000

Email address: `yarmola@math.princeton.edu`

Index

SELECTED PUBLISHED TITLES IN THIS SERIES

79 **Maria Trnkova and Andrew Yarmola, Editors,** 3D Printing in Mathematics, 2023
78 **François Delarue, Editor,** Mean Field Games, 2021
77 **Pablo A. Parrilo and Rekha R. Thomas, Editors,** Sum of Squares: Theory and Applications, 2020
76 **Keenan Crane, Editor,** An Excursion Through Discrete Differential Geometry, 2020
75 **Michael Damron, Firas Rassoul-Agha, and Timo Seppäläinen, Editors,** Random Growth Models, 2018
74 **Jan Bouwe van den Berg and Jean-Philippe Lessard, Editors,** Rigorous Numerics in Dynamics, 2018
73 **Kasso A. Okoudjou, Editor,** Finite Frame Theory, 2016
72 **Van H. Vu, Editor,** Modern Aspects of Random Matrix Theory, 2014
71 **Samson Abramsky and Michael Mislove, Editors,** Mathematical Foundations of Information Flow, 2012
70 **Afra Zomorodian, Editor,** Advances in Applied and Computational Topology, 2012
69 **Karl Sigmund, Editor,** Evolutionary Game Dynamics, 2011
68 **Samuel J. Lomonaco, Jr., Editor,** Quantum Information Science and Its Contributions to Mathematics, 2010
67 **Eitan Tadmor, Jian-Guo Liu, and Athanasios E. Tzavaras, Editors,** Hyperbolic Problems: Theory, Numerics and Applications, 2009
66 **Dorothy Buck and Erica Flapan, Editors,** Applications of Knot Theory, 2009
65 **L. L. Bonilla, A. Carpio, J. M. Vega, and S. Venakides, Editors,** Recent Advances in Nonlinear Partial Differential Equations and Applications, 2007
64 **Reinhard C. Laubenbacher, Editor,** Modeling and Simulation of Biological Networks, 2007
63 **Gestur Ólafsson and Eric Todd Quinto, Editors,** The Radon Transform, Inverse Problems, and Tomography, 2006
62 **Paul Garrett and Daniel Lieman, Editors,** Public-Key Cryptography, 2005
61 **Serkan Hoşten, Jon Lee, and Rekha R. Thomas, Editors,** Trends in Optimization, 2004
60 **Susan G. Williams, Editor,** Symbolic Dynamics and its Applications, 2004
59 **James Sneyd, Editor,** An Introduction to Mathematical Modeling in Physiology, Cell Biology, and Immunology, 2002
58 **Samuel J. Lomonaco, Jr., Editor,** Quantum Computation, 2002
57 **David C. Heath and Glen Swindle, Editors,** Introduction to Mathematical Finance, 1999
56 **Jane Cronin and Robert E. O'Malley, Jr., Editors,** Analyzing Multiscale Phenomena Using Singular Perturbation Methods, 1999
55 **Frederick Hoffman, Editor,** Mathematical Aspects of Artificial Intelligence, 1998
54 **Renato Spigler and Stephanos Venakides, Editors,** Recent Advances in Partial Differential Equations, Venice 1996, 1998
53 **David A. Cox and Bernd Sturmfels, Editors,** Applications of Computational Algebraic Geometry, 1998
52 **V. Mandrekar and P. R.Masani, Editors,** Proceedings of the Norbert Wiener Centenary Congress, 1994, 1997
51 **Louis H. Kauffman, Editor,** The Interface of Knots and Physics, 1996
50 **Robert Calderbank, Editor,** Different Aspects of Coding Theory, 1995

For a complete list of titles in this series, visit the
AMS Bookstore at **www.ams.org/bookstore/psapmseries/**.